DATE DUE

The Long Tomorrow

THE LONG TOMORROW

*How Advances in Evolutionary Biology
Can Help Us Postpone Aging*

Michael R. Rose

OXFORD
UNIVERSITY PRESS

2005

OXFORD
UNIVERSITY PRESS

Oxford University Press, Inc., publishes works that further
Oxford University's objective of excellence
in research, scholarship, and education.

Oxford New York
Auckland   Cape Town   Dar es Salaam   Hong Kong   Karachi
Kuala Lumpur   Madrid   Melbourne   Mexico City   Nairobi
New Delhi   Shanghai   Taipei   Toronto

With offices in
Argentina   Austria   Brazil   Chile   Czech Republic   France   Greece
Guatemala   Hungary   Italy   Japan   Poland   Portugal   Singapore
South Korea   Switzerland   Thailand   Turkey   Ukraine   Vietnam

Published by Oxford University Press, Inc.
198 Madison Avenue, New York, New York 10016
www.oup.com

Oxford is a registered trademark of Oxford University Press

Library of Congress Cataloging-in-Publication Data
Rose, Michael R. (Michael Robertson), 1955–
The long tomorrow : how advances in evolutionary biology can
help us postpone aging / Michael R. Rose.
p. cm.
Includes bibliographical references and index.
ISBN-13: 978-0-19-517939-2
ISBN-10: 0-19-517939-0
1. Aging. 2. Evolution. 3. Longevity. I. Title.
QP86.R593 2005
612.6'7—dc22
2005001820

1 3 5 7 9 8 6 4 2
Printed in the United States of America
on acid-free paper

To Brian Charlesworth

Contents

Foreword

Andrew Weil, M.D.

The possibility of extending human lifespan raises questions that divide the scientific and medical communities. Many bio-gerontologists contend that our lifespan is more or less fixed at about 120 years. They argue that genes for death or aging do not exist, because natural selection can only operate up to the time that organisms reproduce. But researchers have dramatically extended the lifespans of yeasts and worms by identifying and manipulating master regulatory genes that control metabolism. Meanwhile the siren call of anti-aging medicine attracts more and more consumers. Practitioners claim to have their hands on modern versions of the fountain of youth in the form of human growth hormone, antioxidant cocktails, and a variety of products and services not endorsed by their conventional colleagues.

It is not easy to steer a sensible course through all the claims and counterclaims, especially because the subject is an emotional one that touches on our deepest hopes and fears. All of us would like to avoid or at least postpone the loss of independence and general decline that accompany aging, and all of us would like to push back the time of death.

In this book, Michael Rose successfully navigates through the confusion. An evolutionary biologist who has enabled fruit flies to live beyond their normal limits through selective breeding, he comes to these conclusions: 1) using evolutionary mechanisms there is no theoretical barrier to the extension of human lifespan and postponement of aging; 2) at present no practical methods of achieving those goals are available; and 3) a concerted and vigorous scientific effort—he likens it to a biological Manhattan Project—might produce results within a few decades.

I agree. As a physician interested in healthy aging, I am not concerned with stopping or reversing the aging process or even trying to slow it down in myself or in my patients. I am convinced that aging and age-related disease are not synonymous, that it is possible here and now to increase your chances of having a relatively long, healthy, and enjoyable life, with a short and rapid period of decline at the end. The technical term for this is "compression of morbidity," which means squeezing the inevitable decline at the end of life into the shortest amount of time you can.

A first step toward healthy aging is acceptance of the aging process rather than denial of it. Obsession with anti-aging and with life extension can impede that acceptance and lead people to pursue strategies and therapies that may be counterproductive. Given the prominence of anti-aging medicine and the explosion of knowledge in aging research, it is important to have basic knowledge of what we do and do not understand about growing old and what we can and cannot do to influence it.

I trust Michael Rose as an expert guide in this territory.

Tucson, Arizona
April 2005

Preface

In this book I explain how aging is ultimately controlled by evolution, a finding that is full of promise for human welfare. I tell this story from my own particular perspective, as it developed over the last 30 years. Sometimes I mention personal events in my life, but this is not a memoir. Scientific colleagues should be warned that this is not a technical work. I have published several academic books on aging: *Evolutionary Biology of Aging*, *Genetics and Evolution of Aging* (with Caleb Finch), and *Methuselah Flies: A Case Study in the Evolution of Aging* (with Hardip Passananti and Margarida Matos). (Consult the bibliographic essay at the end of this book for publication details and comments about individual references.) Those books supply the fine detail, the thousands of results, suitable for consumption by those from academic science. This book is meant to be read by most anyone, with pleasure when I can turn a felicitous phrase, at least with interest when I can't. In other words, the book you are now examining is intended to be accessible to thoughtful readers who took college courses in Roman history or beat poets, maybe a nonmajor's class in biology.

It is always difficult to explain science. Good science uses mathematics as its language and its theoretical foundation. My work has been based directly on mathematical theory, and the experiments done in my lab have only worked because my students and I have used very large amounts of numerical data. There are no cute pictures from my lab. What we do is hard-core and exacting, not pretty or sweet. It is more in the spirit of building rockets and atomic power plants than it is like cell biology, although genomics and proteomics are slowly pushing cell biology, the specialty of the math-phobic, toward some appreciation of quantitative thinking. Nonetheless, this book can't be quantitative, and as such it isn't "the real thing." If I lapse into telling what are sometimes called "just-so stories," it is only because I have no desire to inflict partial differentiation on my present readers. I have, however, done just that on previous occasions. There is nothing "just-so" about serious evolutionary research on aging.

What I offer here is an intuitive palimpsest of the process by which the evolutionary controls of aging were worked out. I also try to give some feeling for the optimism that is spreading hesitantly, sometimes under threat of institutional persecution and certainly in the face of intimidation, among biologists who study aging for a living. I have spent the better part of three decades postponing aging in fruit flies. Though I started with little faith in the prospect for understanding, much less controlling, aging, my years of work have accumulated into a case for the feasibility of postponing human aging, slowly but substantially, starting from an evolutionary foundation. One of the ubiquitous dreams of mankind, the conquest of our aging is now forseeable. It hasn't arrived yet, but a juggernaut has begun to roll forward, and it is picking up speed. I doubt that the scientific campaign to increase the time span of human health can be stopped. Slowed, yes, it could be. But a long to-morrow is coming, inevitably.

—Irvine, January 2005

The Long Tomorrow

I

The Sphinx and the Rabbi

Early one Sunday in March 2000, I gave a lecture standing next to a sphinx. My audience was likewise surrounded by Egyptian artifacts, some stone panels displaying hieroglyphics: Anubis, Horus, ankh. We were meeting in the University of Pennsylvania's museum, the Egyptian room specifically. It was a relatively formal occasion, so the men were mostly in suits, the women in dresses. The sphinx faced this Sunday congregation with utter indifference. It was the biggest thing in the room, perhaps ten feet tall and at least fifteen feet long, with the head of a pharaoh and the body of a lion.

I was there alongside my new Egyptian friend to speak about biological aging for a Templeton Foundation symposium called "Extended Life, Eternal Life." The Templeton Foundation is interested in the relationship between science and religion. They put on meetings about God and physics, the existence of God, and so on. Our meeting was somewhat less exalted. We were there to bring the science of life span together with the concerns of ethicists and theologians.

It was my exigent task to start the meeting by introducing the science of aging. I began by saying that the ultimate scientific question is why we age. The practical challenge is postponing human aging, if not eliminating it. I offered the opinion that evolutionary biologists had found a good explanation of aging. We had discovered methods for slowing aging. It had even been found that aging can stop completely. I demonstrated all this to the audience using the little red-eyed fruit fly, *Drosophila*.

Some of my biological colleagues followed up with their work. Robert Arking gave a talk about fruit fly experiments of his that were closely related to mine. It was pleasant, and I particularly enjoyed the archeological artifacts. You never got props like that at scientific meetings.

I had a catered lunch with Nicholas Wade of the *New York Times*, seated beneath a vast dome. He is one of our best science writers, and was certainly thoughtful over our meal. Wade expressed considerable un-easiness about the wilder tendencies of the anti-aging movement. Cloning of humans, stem-cell research, where was it all going? Did I really think that we could attain anything like biological immortality? I had the feeling that he wasn't entirely sure why he had come to the Templeton meeting, a feeling that I shared at that moment.

My response was that the problem of aging had always attracted interest from fringe elements. Indeed, whole religions had been erected out of the perennial human aspiration to control aging. Just looking around the museum illustrated this propensity of our species, particularly the elaborate cosmology of reincarnation created by the ancient Egyptians. But this time, I said, we have the prospect of doing something concrete and sensible about aging, whatever the surrounding hysterics. The evolutionary analysis of aging, I contended, was as solid a piece of science as anything in physics. We wouldn't be able to postpone human aging substantially right away, but we could probably make some progress fairly soon. In keeping with the tone of my morning talk, I was trying to be reasonable and dispassionate. Wade seemed to have the same inclination.

That tone was completely undermined after lunch with the speech of Leonard Hayflick. While Hayflick is justly famous for discovering limits to the survival of human cells growing in glass vessels, he is almost as well known for opposing efforts to significantly increase human life span.

That day was to be no exception. Hayflick came out berating those who would alter human aging, calling work to postpone human aging "fundamentally flawed," even snake oil. He referred to this effort as flying in the face of biological facts. He predicted that human aging couldn't be fundamentally improved. It was a practical impossibility, he contended. Then he argued against the ethics of substantially increasing the human life span. I wondered, why did he argue against efforts to achieve something that he said was an impossibility? If human aging can never be delayed, there is no ethical issue. Finally, he said aging was to be "viewed as a blessing" because it killed off the world's undesirables.

After Hayflick, the theologians spoke. Leading off was Diogenes Allen, a Christian philosopher from the Princeton Theological Seminary, a tall, bespectacled man. He began by talking about Perfect Love, which he described as a uniquely blessed state that could be attained only transiently during our time on earth. He quoted descriptions of this state from Dante and W. H. Auden. As far as Allen was concerned, the meaning of human existence was the attainment of that state.

In Heaven, he was confident, we could know an eternity of Perfect Love in God's embrace. In this sense, he argued, death is a consummation of Perfect Love. Our fulfillment, our purpose, would come only at that time. Everything before death would be, at best, only a journey toward God's embrace. Therefore, Allen concluded, "It is a blessing that this life is of limited duration." Those who would delay death only show a "craven fear" that seeks to perpetuate a "selfish life."

Sitting on my uncomfortable chair in the audience, I found this talk as difficult to accept as Hayflick's. I wondered about Buddhists who might never be welcomed into Heaven by the Christian God. They might want to put off the moment of rejection for as long as possible. Some might feel that they have important work to do in this life. Still others might not believe in gods at all. But these ruminations didn't prepare me for what happened next.

The following speaker was Neil Gillman, an eminent rabbi from the Jewish Theological Seminary of America, as well as the Aaron Rabinowitz and Simon H. Rifkind Professor of Jewish Philosophy. I was very curious to hear him speak, because I had never heard a rabbi speak at a conference on aging. Christian theologians, yes, but never a Jewish philosopher. Short, with a little white goatee, Gillman spoke as if he were reciting poetry. He had no interest in arguing against Christian values. But he pointed out that, in Judaism, life was good. "Judaism understands God as life." That which preserved life was good. "Sex is good. Food is good."

I had never heard anything like this before. But then I knew little about Jewish theology. I wondered just how far Gillman would go. The Christian theologians were shifting uncomfortably. The tables were about to be turned, if not the tablets.

The rabbi drove on. "There is nothing redemptive about death. Death is absurd. Death is my enemy." And if life is good, the extension of life must be good. More life is better than less life, he said. After his talk, he was asked by a Christian theologian whether he really believed in the extension of healthy life. Yes, Gillman replied. If we could live 20 years longer, they asked, would it be good to live still another 20 years? Yes, yes, he said. And then he apologized for having to leave, because he had another appointment in Queens. This spared him from hearing his fellow theologians refer to views such as his as a "sub-biblical plague," showing "an ignoble fear of death."

3

A few things had become obvious to me. Biologists didn't agree about the desirability of extending the human life span. Theologians didn't agree about it either. But I was also certain that the entire issue of human aging, its science, its postponement, and its meaning had moved toward the center stage of modern society. For all the print that is given over to AIDS, and rightly so, aging is still the most common factor in the deaths of the majority of adults in industrialized countries. And now evolutionary biology had given us a solid scientific understanding with which to base attempts to control aging. Before, there had been too many unanswered questions. Could the limits of a species life span be transgressed? How fast could aging be changed? Was there anything special about aging in humans? Now, we have answers for all these questions and more.

For me personally, it has been intellectually thrilling to answer some of these questions about the foundations of gerontology, the scientific study of aging. The progress in the evolutionary biology of aging has been momentous, both scientifically sweeping and medically inspiring. As the first full-time evolutionary gerontologist, I had deliberately extended the length of life beyond its normal limits. It is true that I did this just with the humble laboratory fruit fly. But much of modern biology is based on experiments performed first with fruit flies. That initial work on the evolution of aging in fruit flies has radiated outward, to other experimental animals, and to plans for slowing human aging. That is my story.

But it doesn't begin quite the way you might suppose.

2

Maynard Smith's Shirts

My entry into the field of aging research was entirely accidental. I fell into it almost the way Victorian Alice fell through the white rabbit's bolt hole. In my case, John Maynard Smith was the white rabbit that I pursued and the year was 1975.

On the last day of a biology conference that summer of 1975, I knocked on the door of Maynard Smith's room. He was then one of England's greatest biologists, one of the few who hadn't left for more money in America or better weather in Australia. At that moment, there was no one in the world I admired or respected more. I might have been knocking on God's door, for all the determination it took.

There was no answer. I knocked again. Still no answer.

I began to grow nervous. I had spent two weeks trying to make a good impression on Maynard Smith at the meeting, which was called "Mathematics and the Life Sciences." The location was Sherbrooke, Quebec, not far from the town of Sorel, where my family had lived in the nineteenth century. The midsummer weather had been glorious, sunny but never stifling, and the conference very much to my liking. I was interested in becoming a theoretical biologist. Maynard Smith, a professor and former dean at the University of Sussex, was one of the world's leading theoretical biologists. I had read his scientific articles before coming to the meeting, but they hadn't prepared me for his verbal powers. John could captivate an audience of scientists like Elvis Presley singing to a Las Vegas crowd. I desperately wanted to work with him for my doctoral studies.

The door finally opened. Here was Maynard Smith, long gray hair swept up and back, pigeon-eyed glasses, skin with patches of red and white. The great man was out of sorts.

"I'm afraid we can't talk now," he said. "Something's come up." He seemed to be distressed that I had appeared at his door, even though we had made an appointment.

The last two weeks swam before my eyes. Was it my refusal to drink beer with him? My lack of English manners? Probably, I thought, it was when I asked him how old he was, because I was afraid that he would retire soon. He had laughed, and asked how old he looked. I had blurted, "About 80," and he didn't laugh anymore. He was 55 at the time.

Perhaps noticing my dismay, Maynard Smith relented and invited me into his room, which was in disarray, clothes everywhere. "It seems I've forgotten how to fold shirts," he said in his best Peter O'Toole style.

I was relieved. I had been folding shirts all my life, thanks to an itinerant existence as a military brat in Europe and North America. I set about folding Maynard Smith's shirts while we talked.

I made it clear to him that I had found the last two weeks of talks on mathematical biology, and especially his lectures, quite exciting. I made particular mention of the elegance of his work in evolutionary game theory. I told him that I wanted to come to England to work as his graduate student. To be more accurate, I begged.

After some polite digression to soften the blow, Maynard Smith said that he had grown tired of the duties required to supervise graduate students, preferring to work on his own, sometimes with faculty colleagues. I replied that I wouldn't need much baby-sitting, because I had already been doing research for two years. But Maynard Smith wouldn't give yes for an answer.

Instead, he pointedly brought up Brian Charlesworth, a new faculty member he had just brought to the University of Sussex. As a new member of their faculty, Charlesworth needed graduate students. I knew that Charlesworth had already published some good theoretical papers, but he really wasn't the man I wanted to work with.

"Of course, you will have to do a lab project," Maynard Smith went on, oblivious to my lack of enthusiasm, or at least unheeding. "We aren't really interested in students who just do maths. Brian has a project on aging."

This I didn't like. I wanted to do mathematics, theoretical biology. That had been the whole appeal of Maynard Smith, and of the conference. Solving grand theoretical problems using a pen and paper. I had no taste for experimental work, with its tedious collection of a few pieces of data a day. There wasn't enough of an intellectual thrill to experimental work, I thought. And who cared about aging, anyway, I thought to myself. I had just turned 20 and naturally had no emotional conviction that I would ever grow old myself.

Disheartened, I left Maynard Smith to the rest of his packing.

I have spent most of my life working on aging despite my initial misgivings. While I had been attracted to his theoretical work on sex and social behavior, Maynard Smith first made an impact on the world with his research on aging in fruit flies. Starting in the 1950s at University College London, he had been one of three great experimentalists of that era who worked on aging. All three were born in the early 1920s, the other two being Alex Comfort and Roy Walford—both of whom you will meet later. All three died in the last five years, victims of their own research topic, Walford and Maynard Smith just weeks apart in 2004. They were aging's three wise men, present at the birth of the field's scientific respectability, and responsible for that birth in many ways.

Maynard Smith and his peers changed aging from a field dominated by badly conducted research to a field in which PhD's could work respectably. At first they made little progress. While I still cite Maynard Smith's work on aging from the 1950s and 1960s, few biologists publishing now have even heard of it. They think of Maynard Smith as a great theorist of biology, especially his game theory. My opinion is that he was such a great theoretician partly because he had actually done experiments with living things, rather than being a mathematician or theoretical physicist who slummed in biology to pick up easy publications. And those experiments, like many of the ones I will tell you about here, used humble, tiny fruit flies. There is nothing odd about that in biology. Thomas Hunt Morgan and his students founded modern genetics using fruit flies. Neurobiologists, developmental biologists, and evolutionists have followed their example, making the laboratory fruit fly one of the best-known animals in the scientific universe.

It would be a good story to say that I foresaw my future work on aging in a blinding flash on leaving John Maynard Smith's room that morning in the summer of 1975, but it wouldn't be true. It took me a long time to get over my fascination with theory, to see the value in working in a lab using a tiny yellow insect with red eyes.

My beginnings as a scientist weren't very auspicious. My first experiment took place in Ottawa when I was seven years old, sitting on a bench between the back door to the family house and our garage. It was the ferocious Canadian winter, snow and ice everywhere, and I was bundled up to go skating. My skates had been laced up in the kitchen and I would wear them during the ride to the ice rink. As I waited for my mother to get ready, I wondered how ice skates worked. Did the blade cut directly into the ice like a knife, or did skates work by cutting into the ice with their edges? If the first theory were correct, the edges of the blade should be round or smooth, so that the middle of the blade came to a point. Seeking to test this theory, I set about "sharpening" my skates on the concrete walk in front of me, scraping the blades sideways to round their edges. I spent the rest of the day falling down on the ice. The first theory was wrong. The edges of the blade had to be sharp, because they cut into the ice at an oblique angle when you push on the skate, supporting the second theory. I never told anyone this story before writing the present passage, perhaps because I was embarrassed that a Canadian boy had to do such an experiment. After all, we are given our first pair of ice skates in the delivery room.

That same year my great-aunt Charlotte sent me a little microscope, the real thing with two lenses and metal parts packaged in a wooden box. I couldn't have been more excited. My parents took one look at the marvelous device, complete with glass slides and sharp tools, and took it away from their young son. I was heartbroken; perhaps I cried. I got the microscope back in a year or so, but by then it was no longer such a thing of wonder. In protecting me from my juvenile curiosity, my parents probably spared me numerous small cuts and a career as a pathologist.

By way of compensation perhaps, I became interested in astronomy. A friend of mine had a reflecting telescope that was much superior to my refracting scope, and the moon was an intensely luminous white in its visual field. But the real excitement for me was the theories of physics and astronomy. There was the famous controversy about the geocentric theory of the universe versus the Copernican heliocentric solar system. Like many young people since 1700, I first heard about the relationship between scientific theory and experiment from that story. In particular, I learned that the correct scientific theory may not be the theory that everyone thinks is obviously true, at least at first. This taught me to be wary of a theory if it is only plausible, before direct tests have been made.

I was also a fan of Fred Hoyle's steady-state theory of the universe, even though it was to be demolished by the rival "big bang" theory in the years to come. Cosmology was my favorite topic, though I had no one to discuss it with as a child. I remember pleading with my father in a Montreal bookstore when I was 11; I wanted to buy a cosmology paperback called, to the best of my recollection, *The Oscillating Universe*. It cost about 50 cents, back then in 1966, but I had to borrow against my allowance to buy it. My father couldn't imagine why I wanted the thing.

My infatuation with scientific theory continued into my undergraduate years. I was lucky enough to meet Dolf Harmsen, a biology professor at Queen's University, in Ontario, where I got both bachelor's and master's degrees. A genial Dutch-Canadian, Dolf let me indulge my interest in theory despite his background in insect physiology, one of the most obdurately experimental fields in biology. He had a taste for theory himself. Under Dolf's patient direction, I created computer models of the ecology of the forest tent caterpillar, a species that can denude entire forests when it is having a population outbreak. He supported my dream of becoming a theoretical biologist when I was just 18 years old, wore wild clothing, and had a reputation for eccentricity so bad that it still endures at Queen's. My debt to Dolf is enormous. I might have ended my years running a record store if he hadn't let me have a desk in his lab and a computer account. On the other hand, I might have made more money with the record store, like Richard Branson did with Virgin Records—or so I fantasize.

After being rejected by Maynard Smith, I pursued a backup plan, and a good one I thought, to work with Harvard's Richard Levins and Richard Lewontin. Levins and Lewontin were two of the leaders of left-wing biology, a style of science that had been very popular in the 1930s and was still clinging to life in the 1970s. Richard Levins had been at the 1975 Sherbrooke conference too. While his lectures hadn't had the range of Maynard Smith's, they had still been intriguing. Besides, Levins was enormously charismatic, a cross between Fidel Castro and Ernest Hemingway, built like a bull, with rabbinical receding hair. He told interesting stories about his experiences in communist countries, especially Cuba, where he had of course helped with the sugar-cane harvest. As I had been a party worker for Canada's socialist New Democratic Party, I pictured Cuba as a little third world utopia free of the malign influence of Western capitalism. This was 1975, after all.

That fall I went to Harvard to meet Levins's comrade, Richard Lewontin. The meeting had been scheduled some weeks in advance, and it had not been noticed that it fell near American Thanksgiving. As a Canadian, I didn't have a clue about American Thanksgiving. When

I showed up that week, accompanied by my father, the graduate students in Lewontin's lab were greatly amused. He was summoned by telephone from his home in Vermont. It was kind of him to come back to Cambridge just to meet me during his Thanksgiving.

In 1975, Richard Lewontin was well known for his flair and his accomplishments. Preternaturally boyish in his forties, with a shock of black hair and owlish glasses, he was at the top of his game. His achievements and opinions had dominated the field of evolutionary genetics since the 1960s. Among other things, Lewontin made evolutionary biology molecular. He was also mathematically talented and a writer of considerable gifts.

The project that Lewontin and Levins offered for my PhD at Harvard was to build computer models of the disease schistosomiasis, a notable scourge of tropical countries. Schistosomes are parasitic worms that inhabit internal organs. More than 200 million people are infected with these flatworms, and about 800,000 die of the disease each year. Helping to control this disease would have been a significant humanitarian achievement. It would also have been a straightforward continuation of the computer work that I was doing with Dolf Harmsen, and everyone was confident that I could do the job, especially me. Maybe I was too confident; the project didn't seem like much of an intellectual challenge at the time.

As 1975 became 1976, I was leaning toward Harvard, but still curious about the possibility of working with Brian Charlesworth at the University of Sussex. It did, after all, offer the prospect of being near Maynard Smith, and keeping alive my dreams of doing grand biological theory. At that time, Charlesworth was little known in North America. He had in fact been one of Lewontin's protégés. From 1969 to 1971, he had worked as Lewontin's postdoctoral student at the University of Chicago, before Lewontin moved to Harvard. In 1976, Brian was about 30 years old. He had published little experimental work, though he had produced several important mathematical articles on the connection between ecology and evolution. This was a topic I found extremely interesting. Indeed, it was one of the hottest themes in biology at that time.

Then, as now, the contrast between Harvard and Sussex could hardly be overstated. Harvard has the greatest single university campus in the world, even if the University of California as a whole is ahead of it in almost every other respect. Harvard's Nobel laureates and disciplinary leaders speak for themselves, literally as well as figuratively. Founded in the 1600s, older than most European universities, Harvard had once offered Galileo a faculty position.

In the spring of 1976, I went to Harvard to see Lewontin and Levins for a week. Wandering around Harvard "Coop" and the bookstores of "Cambridge Mass" on my own was a heady experience for an aspiring scientist. The abundance of street people and police bearing side arms was a bit disconcerting for a young Canadian, but back then I did not understand how different American cities were from those of Canada.

As it happened, my visit came at the peak of Harvard's sociobiology controversy. *Sociobiology* was E. O. Wilson's 1975 book that attempted to unify biology with the social sciences, using evolutionary theory as their common foundation. Notable among Wilson's claims was the idea that we would soon be able to unravel the mysteries of human behavior using sociobiology. This claim outraged those with doctrinal links to the traditional social sciences and various political and theological ideologies. The ensuing controversy was of such interest to the general public, or at least to journalists, that it was a common cover story in the magazines of 1976. E. O. Wilson was a Harvard professor of zoology, but his efforts were strongly resisted by other Harvard professors, most notably by Richard Lewontin, whose lab was in the same building as Wilson's. E. O. Wilson had been one of the primary proponents of Lewontin's recruitment to the Harvard faculty, and they had many scientific interests in common. In some ways the situation was reminiscent of Shakespeare's *Julius Caesar*, intellectual fratricide among the Harvard elite paralleling assassination among the Roman elite.

One of the first things that Lewontin did to counter Wilson was form the Sociobiology Study Group. That spring, I attended the climactic meeting of the study group in Lewontin's vintage Cambridge apartment, the meeting where they discussed the last chapter of *Sociobiology*, the one on human behavior. That was the stuff they all hated, the part where Wilson announced that sociobiology would replace the traditional social sciences. I had great expectations for this meeting.

The living room was full of Harvard luminaries from a range of disciplines. Stephen Jay Gould was there, grinning throughout, not yet a famous popular science writer and controversialist. We graduate students mostly sat on the floor, at the feet of the great professors who occupied a miscellany of chairs, stools, and sofas. The stage seemed to be set for spectacular fulminations.

Instead, the meeting began with an exercise in Maoist constructive criticism. A few of the graduate students lambasted Lewontin and the other professors for their elitism and sexism, their neglect of the opinions of the graduate students, particularly the female graduate students. Lewontin seemed to be profoundly embarrassed by this, squirming in his chair, squinting behind his glasses. Yet his Marxist manners required that he confess to these sins of sexism and elitism. At close range, his self-abasement was horrible to watch.

After some time the meeting returned to the matter at hand, denigrating Wilson. But after the opening contretemps, there was little enthusiasm for the task. It was generally agreed that Wilson's views were totally unacceptable. But the scientific level of the discussion was shallow. A few of the more egregious examples of reductive thinking in his book were held up for ridicule. I offered one such example, though I can't remember what it was almost 30 years later. Evidently my comment wasn't as insightful as I thought at the time. Then the meeting disbanded, a great let down, at least for me.

Brian Charlesworth sent me a handwritten letter early in 1976, proposing that I work under him on the evolution of aging. Maynard Smith had told him about me. It was the research assignment I just didn't want. Brian argued in his letter that evolutionary theory had solved the problem of aging. All he needed now was a graduate student who could show this experimentally. He was offering me the chance to address one of the deepest problems in all of biology.

What was I to do? The academically glamorous tumult of Harvard or Charlesworth's project on aging?

3

Cell Gang

If I had been Promethean, I would have embraced the challenge that Brian offered at once, never mind the little detail that I didn't understand what Charlesworth was talking about in his letter. Solve the mystery of aging using evolutionary theory? How?

I wrote back to him expressing considerable doubt about anyone solving the general problem of aging, least of all me. Surely the specific ecology and physiology of each species would determine its aging, I suggested, not some general evolutionary mechanism. I also expressed the prudential sentiment that I didn't want to ruin my career by pursuing a risky line of research.

Looking back on this episode decades later, I ask myself why didn't I see that Brian was offering me a dream project for my doctorate? How many beginning scientists are ever given such an opportunity? From what I had seen, most doctoral students were given the dog's breakfast.

The track record of work on aging did not inspire optimism in me. For most of history, aging hasn't been the study of dispassionate scientists. Instead, it has been the preserve of charlatans and unethical medical doctors. Against this backdrop the scientific study of aging limped along, a Quasimodo regularly beaten for the sins of the charlatans, suffering from their extremely unsavory odor, and starved for funding. Science, like American politics and French painting, is subject to bandwagons, to fashion. When a new research approach has success, it quickly attracts adherents. These new devotees are motivated by several things. One is the inherent interest of the research ideas. Another is the opportunity to do research that works, to develop theory that gives elegant results, to perform experiments that can be interpreted. Not to be neglected is the fact that doing hot research nets scientists publications, research grants, and faculty positions. The scientific study of aging was not a hot area in 1976. There was no bandwagon, not even a minivan. There was a simple explanation for this: failure.

In the first half of the twentieth century, physiologists and medical doctors made a train wreck out of the field of aging. Consider the most eminent biologist to study aging before World War II: the Russian Elie Metchnikoff. He was the winner of the 1908 Nobel Prize and a follower of Louis Pasteur, Pasteur being the scientist who firmly established the germ theory of disease. Metchnikoff shared with his master an obsession with diseases caused by microbes. Metchnikoff generalized this idea to aging, proposing that aging was caused by the toxic effects of bacteria living in the gut, including counterproductive effects of our phagocytes destroying "bad" bacteria. In a sense, he proposed that our *intestines are a microbial battlefield* that caused pathology throughout the body. He bolstered this conclusion by pointing to birds, which have small intestines and live longer than mammals that have larger intestines. He also surgically removed the bowels of Parisian patients, and then argued that they had improved health as a result. Even the patients were convinced. Metchnikoff's theory became quite popular, motivating much general concern about bowel movements, roughage, and the consumption of lactophilic bacteria in fresh yogurt. You can still walk the aisles of grocery stores and see products like breakfast cereals that owe their popularity to the diffusion of Metchnikoff's bowel theory. A rather unpopular movie, *The Road to Wellville* (1994), was made based on the movements he inspired.

Unfortunately for Metchnikoff's theory, and laxative makers generally, later experiments that reared laboratory animals without bacteria in their guts did nothing for their aging. Sometimes they died sooner. But he himself was dead by the time these damning results were published.

It is no surprise that bacteria aren't the cause of aging, at least it is no surprise now. Diseases that are typical of aging in mammals, such as cancer and heart disease, are rarely caused by bacterial infections, we now know. Metchnikoff's explanation of aging was a classic example of an attractive theory killed by an ugly fact. But that's how science works. It's about truth, not beauty. Not even clean bowels.

In fairness, however, I should mention that some of Metchnikoff's ideas have been polished up and resuscitated by biologists like Paul Ewald and Caleb Finch. Finch, in particular, is impressed with the connection between aging and inflammation due to chronic infection. I would like to see more direct experimental evidence bearing on this idea. In any case, these latter-day biological theorists are far more advanced than Metchnikoff.

The first half of the twentieth century saw more theories of aging like Metchnikoff's, proposed by similarly eminent biologists. Each of these

theories was based on an intuitively attractive physiological speculation. To give another example, G. P. Bidder proposed in the 1930s that
aging was caused by the *cessation of growth* in adults. Organisms that
grow without limits, like some trees and some fish, Bidder reasoned,
should have death rates that do not increase with age, the way they do
in most animals and plants. And indeed trees are among the longest-
lived organisms.

Bidder's growth theory of aging was combined with Metchnikoff's
bowel theory in a novel by Aldous Huxley, *After Many a Summer Dies
the Swan* (1939). In Huxley's somewhat satirical story, a dissipated English nobleman overcomes aging by eating the raw guts of carp that
continue to grow as adults and age very slowly. This novel was better
than *The Road to Wellville*, especially some acidic episodes concerning
the relationship between unscrupulous medical doctors and wealthy
patients afraid of death. But I don't want to spoil it for the reader; you
should peruse Huxley's novel for yourself rather than reading what I
think of it.

The sharp-witted Alex Comfort, from whom we will hear more,
demolished Bidder's theory. Comfort studied aging in aquarium fish
that don't stop growing during adulthood. His measure of aging was
the death rate of the fish, relative to their chronological age. The fish
aged despite their continued growth. Bidder's interesting theory was
yet another casualty of the experimental method.

By the middle of the twentieth century, the obvious explanations of
aging had all been tried and found wanting. Whatever the solution to
the aging puzzle was, it wasn't going to be some simple piece of biology. A complex or hidden cause must be acting. It wasn't long before
such hidden causes were proposed.

Everything changed in the middle of the twentieth century for the study
of aging, indeed for the study of biology generally. Molecular and cell
biologists began to take over biology, starting with future Nobel laureates James Watson and Francis Crick publishing the double-helix model
for DNA in 1953, and the demonstration that DNA molecules encode
genes. Molecular and cell biologists have owned biology ever since.
This gave them the hubris to try on the problem of aging. Fortunately,
they had some new ideas. Unfortunately, they didn't work any better
than the old ideas.

One of their first efforts was the *somatic mutation* theory of aging,
which flourished from the 1950s to the 1960s. This theory assumes

that aging is caused by genetic mutations in the body's somatic cells, the cells that don't reproduce. ("Soma" means body.) The accumulation of mutations in somatic DNA might, it was thought, screw up our genetic information, leading to the deterioration of the older body, or in other words aging.

The somatic mutation theory grew out of atomic physics in two respects. First, an early proponent was Leo Szilard, the man who first had a practicable idea for how to make an atomic bomb. Second, the effects of atomic radiation seem like accelerated aging. Heavily irradiated mice, for example, just act old: unhealthy and feeble. And then they die. Another attractive feature of the somatic mutation theory is that many things go wrong during normal aging. This made it unlikely that aging could be due to any single process of deterioration, like Metchnikoff's poisoned intestines. Somatic mutations can disrupt health in many ways, because they could, in principle, disrupt the genes responsible for the working of each and every tissue. Somatic mutation could conceivably work as the hidden unitary cause of a complex biology of aging.

Somatic mutation was the most credible, fully developed theory of aging that had been produced as of 1960. But it too was to be destroyed by inconvenient facts. Radiation damage is different from the damage caused by normal aging. It does not give rise to much cardiovascular disease, the way aging does in humans. Cancer, which had first inspired hope for the somatic mutation theory, turned out to be unusual in its strong and parallel connections with both aging and radiation. Indeed, biologists now know that cancer is mostly due to multiple mutations of genes in somatic cells. We even know that radiation often kills because it triggers cancer. But many processes of aging are like cardiovascular disease in that they are not accelerated by radiation. By the end of the 1960s, somatic mutation was dead as a general theory of aging. Somatic mutation is a good explanation of cancer, but it has failed as a complete theory of aging.

The last completely general molecular theory of aging that attracted much attention from scientists was the *error catastrophe theory* proposed by Leslie Orgel in 1963. His starting point was the molecular machinery that cells use to make proteins. This synthetic machinery was well known by 1963. Orgel posed the following question: what if there were errors in the synthesis of the synthetic machinery itself? These errors should make the synthetic process less reliable, leading to still more errors in the synthesis of the machinery. This could in turn cause a progressive accumulation of errors building to catastrophic levels. If organisms run out of correctly made protein, they should develop pervasive pathologies, and thus aging.

This theory was ingenious because it was based on the organization of life, how one key part of the cellular machinery was connected to other parts of the cellular machinery, and back to itself. This type of catastrophe of errors, of being undermined from within, is not particularly hard to understand. Think of the impact of double agents on a nation's counterintelligence service. The people responsible for detecting traitors would themselves be traitors. Under these conditions, the conduct of war and diplomacy would be pervasively undermined due to unchecked infiltration by enemy agents. This happened to British intelligence in the years after 1945, with the infiltration of the double agents Philby, Burgess, Maclean, and Blunt, a conspiracy that grew out of their days as idealistic Cambridge University communists. These men dramatically undermined Western intelligence at the start of the Cold War. In the same way, erroneous synthesis of the body's synthetic machinery would undermine the integrity of the body, perhaps cumulatively. Like the somatic mutation theory of aging, the error catastrophe theory is based on the generation of biochemical errors. Unlike the somatic mutation theory, the error catastrophe theory has an autocatalytic dynamic that takes hold of the cell and drives it toward destruction. This makes it seem much more plausible. Yet the error catastrophe theory shares with the somatic mutation theory the general prediction that the pathologies of aging should be diverse and pervasive, as they indeed seem to be. The error catastrophe theory was easily the most intellectually satisfying of all molecular theories of aging.

There was some experimental evidence that supported the theory, at least indirectly. Genetically altered fungi that always have erroneous protein synthesis suffer deterioration from an increase in defective protein. This shows that error catastrophes happen in mutants with a genetic tendency to errors of synthesis. But do they happen in normal aging, when there are no predisposing mutations?

Many attempts were made to drive normal organisms toward error catastrophes, but none of them worked properly. Maynard Smith and his colleagues tried to make error catastrophes in fruit flies by feeding them chemically aberrant food, food that should have poisoned the protein synthesizing of the fly cells. But it didn't work. Careful study of normal cells has shown that the error rates of their protein synthesis are very low. Normal cells are just really good at making proteins, and making them right, like Lexus makes cars. Therefore, it is unlikely that error catastrophes occur often in cells in nature, leaving aside rare genetic mutants that start life with deficiencies in the accuracy of their protein synthesis. The genetic mutants that aren't good at protein synthesis are like Yugos, that cheap little car they used to make in Yugoslavia when it was still a communist country. (They were especially popular among professors who wanted to show off how stylishly revolutionary they were.) They are bad cars when you first buy them, and they get worse over time. Natural selection eliminates Yugo-style genetic mutations almost immediately, anyway, so they do not play a role in normal aging. The error catastrophe theory is yet another corpse littering the scientific field of aging, like Yugos in a junkyard.

Of more lasting influence was the work of Leonard Hayflick and his cell gang, even though it wasn't theoretically inspired. By the 1970s, Leonard Hayflick had convinced a substantial number of biologists that aging was caused by problems with cell division. While he brought evolution into his reasoning, the fundamental theory and findings came from cell research. This was credible to the biologists of the late twentieth century, because many of the ideas of biology are based on cell biology. Why not aging? After all, cells are literally the building blocks of tissues and organs, the compartments of plasm within which most of the business of the organism was conducted.

Before Hayflick, the consensus within biology had been that aging was due to a breakdown of the physiology of the entire organism. This consensus had its roots in the work of Alexis Carrel. Carrel was an eminent biologist in the first part of the twentieth century, a Nobel laureate. Yes, another one. Carrel came to the United States from Lyon, France, where he had pioneered the use of fine silk ligature in surgery. He was virtually the Victor Frankenstein of the early twentieth century, swapping limbs between dogs by first amputating the legs of two dogs and then the sewing each leg onto the other dog's body. But Carrel gave up on transplantation. Even when his operations were perfect in their sewing, organs that were switched between mammals were destroyed by tissue rejection. These were the days before effective chemical suppression of immune response with cyclosporine.

Among the problems that Carrel worked on, after he gave up on transplantation, was the culture of vertebrate cells in glass vessels ("in vitro"). He was famous for continuously culturing chick cells for many years without interruption. This was achieved by supplying the chick cells with purified chicken serum, an extract from chicken blood.

There is nothing unusual about keeping cancer cells alive for a long time. Even in the early years of the twentieth century it wasn't that hard to keep tumorous cells alive indefinitely. Cancer cells seem to have a remorseless determination to live, to divide, and to spread. Tumor cells had been kept alive indefinitely in experimental animals using repeated transplantation. But no one had been able to indefinitely propagate the cells of vertebrates when they weren't cancerous, before Carrel.

Carrel's lab was unique in its ability to keep normal cells alive. This very uniqueness should have tipped off biologists that something was wrong with this result, but Carrel had a huge reputation in the research community. Just the results from his lab alone had a considerable impact on aging research. In 1922's *The Biology of Death*, Raymond Pearl assumed that the cells of the body have immortality, with the ability to live outside the body far longer than the body lives, under the right conditions. His assumption was that Carrel had shown that a breakdown in the *functional organization* of the body causes aging. Cell aging didn't occur, Pearl thought, so it couldn't explain the aging of the whole body. This idea was popularized in the "character" Chicken Little, a huge culture of chicken muscle in the 1952 novel *The Space Merchants*, by Frederik Kohl and C. M. Kornbluth. Chicken Little was a plentiful source of cheap protein in a future society run by exploitative corporations that marketed dubious food products—nothing like present-day America of course.

Inspired by Carrel's finding, Raymond Pearl pursued the genetics and physiology of aging using *Drosophila*, a forerunner for the work on the genetics of aging that really got going in the 1980s. While Pearl did some interesting experiments on a variety of fruit fly mutants, he never used formal evolutionary analysis to design or interpret his experiments. Deprived of a well-defined theoretical foundation, Pearl's work on aging has had relatively little influence since 1950.

In the late 1950s, Leonard Hayflick wanted to create an immortal, but normal, cell line to produce a lot of cells for medical purposes. In other words, he wanted to create a massive clone of human tissue that could be used endlessly, a Human Little for the purposes of medical research. He had no interest in aging to begin with. But every time he tried to keep a population of normal cells alive in vitro, they eventually stopped growing in numbers. This raised the question, do animal cells age in vitro? And if they do, what happened in Carrel's cell cultures?

Hayflick's interpretation of Carrel's work was that the chicken serum used to feed Carrel's cells was contaminated with new chick cells.
This may have occurred because of a failure to extract cell-free serum
from chicken blood in the first place. The equipment that they used
back then to purify blood products was still fairly crude. (I have also
heard rumors that Carrel's lab assistants performed some of their work
by torchlight wearing hoods, but I have never seen these stories in print.
Biologists just don't admit to being interesting, except perhaps for Jim
Watson of DNA fame.) If Hayflick is right, Carrel's seemingly immortal cells may have instead been a culture of young cells replenished
repeatedly by cells from the intermittently refreshed nutritive serum.

What really happens when present-day cell biologists grow tissues in
culture vessels, away from the bodies in which they normally live? In
order to create an in vitro culture, cells are extracted from living tissue
and placed in glass or plastic culture vessels. These culture vessels are
kept in incubators that have their thermostats set near blood temperature. The culture vessels contain nutritive cell-free serum derived from
blood, to which hormones are added. Cells attach to the surface of the
culture vessel and begin to divide. This is how cell cultures are initi-
ated. No lightning or hunchback assistants are required, though they
do add to the atmosphere.

A period of *rapid cell division* follows. The cells divide until they have
formed a mat covering the vessel surface. Once they have covered the
vessel surface, they stop dividing. The tissue is then cut in half, and one
half is discarded. The remaining cells then resume division, filling up
their culture vessel again, cell number doubling. One of these cycles of
doubling is called a "passage." The period of rapid cell division lasts 50
to 60 passages in cultures started from normal embryonic human cells.

After the period of rapid cell proliferation, the rate of division first
slows and then ceases. At this point, cell cultures are referred to as
moribund or *senescent*. But the cells do not necessarily die off. They
have just ceased active division.

What Hayflick discovered was that every culture of normal human
cells that he started eventually hit this third phase, when the rate of
division decelerates. Cells growing and dividing in the microcosm of
the culture vessel seem to become progressively inhibited about dividing, as if they know that we watch them do it and they're bashful. This
procession of increasing inhibition results in the eventual cessation of
cell division. The number of passages until cell division stops is now

commonly called the Hayflick Limit, a term that I proposed in 1991. I thought he deserved some overt credit for going against the medical establishment of his day.

So it has been shown conclusively that ordinary human cells limit their division. *Why?* One theory is that cells are put together so that they will eventually die, to produce aging of the whole body. This sounds far-fetched, but it might be justified on the grounds that individuals need to age for the good of their family, group, population, or species. Otherwise they might take up space, or use up food.

There are many scientific problems with this group-oriented theory of selection for aging. The most important is that few species show obvious aging in nature. If evolution has created a mechanism to shut down cell division in order to get rid of old animals, the fact that very few animals obviously die of aging in nature is a puzzle. Indeed, aging is rarely detected in studies of wild populations of animals, though it is occasionally seen in nature, one possible example being the rapid deterioration of Pacific salmon after mating. Mostly aging is observed among animals kept under benign conditions contrived by doting humans: pets, zoo animals, and prize specimens of agricultural breeds.

University professors are a good example of organisms that progressively deteriorate while their eternally young audiences watch. Yet evolution could not have conspired to kill these animals, because the benign conditions that they enjoy are recent novelties. Even universities have been around for little more than 30 generations. Evolution hasn't had enough generations to generate aging among these pampered creatures, even if it were trying to kill off older university professors to make way for new PhD graduates among the faculty.

Cancer supplies an alternative explanation as to why somatic cells might normally have limited division. Vertebrate bodies have a lot of cells that divide. Insects, by contrast, have almost no such cells. Insects very rarely get tumors as they age, while mammals often do. Cell division during adulthood is a double-edged sword. Our physiology as adults depends on cells that can divide. Our immune response, for example, requires the production of numerous cells that produce antibodies. But cells that increase in numbers without stopping, without some inhibiting mechanism, might become malignancies that can kill. A molecular mechanism that allows a lot of cell division, but not an unlimited amount, seems like a good idea. Perhaps that is why the Hayflick Limit exists, to control cancer?

Of course that was not the interpretation favored by most of the cell biologists who studied cells in vitro. They believed instead that limited cell division is the premier cause of aging. In support of their opinion they cited studies that indicate a close association between passage number and aging. For example, cells can be cultured for fewer passages as the age of the donor increases, suggesting that aging "uses up" cell divisions.

Even if the age of the body predicts the number of in vitro passages left among the cells extracted from it, this doesn't show that the aging of the body is tied up with the number of cell divisions. A cell that can undergo only 10 more divisions may be able to do its job just as well as one that can go through 50 more divisions. In addition, only a small number of cells in the mammalian body are likely to get to the third phase of cell division, the so-called senescent phase. Few tissues undergo that much cell division. Furthermore cells from older tissues have some remaining capacity to divide. Being 90 doesn't mean that you have run out of cells capable of dividing.

What happens when a cell *does* cease dividing? When those cells are examined closely, they are not obviously moribund. Instead, they vary more in size, shape, and internal structure. This raises the possibility that the termination of cell division that we see in vitro is a prelude to *differentiation* of cells for specific physiological roles, like the differentiation of the specialized cells of the different organs of the body. The active period of cell division in the normal body would, on this interpretation, be a period of quantitative tissue growth, before cells take on specialized forms and functions which require that they stop dividing.

How decrepit are our cells when they have ceased dividing? In the adult human brain, neurons rarely divide. Yet those same neurons function for decades. If a lack of division were a reliable sign that a cell is senescent, neurons would be senescent before the start of adulthood. And that is not remotely plausible, raising a red flag over the assumption that the end of cell division has a close correlation with normal aging.

Hayflick offered an ingenious argument that addressed this point. His idea is that it is not the end of cell division that causes aging, but an underlying process of cell aging that causes both the aging of the whole body and the end of cell division. Hayflick also pointed out that the number of passages that cells manage in vitro is strongly correlated with the maximum life span of a species when it is kept under good conditions. The Galapagos tortoise, which can live well over 100 years, has cells that will undergo 90 to 125 passages. House mice, on the other hand, live at best four years, and their cells can manage just 14 to 28 passages. This is a striking correlation between the number of cell passages and the process of aging. And this finding has been reproduced in several studies. The result seems to tie cell division to aging. It was an interesting speculation of Hayflick's, but it hardly solved the problem of aging. Insects have life spans that vary widely, yet they have little cell division in their adult bodies, making cell proliferation an implausible explanation for their aging patterns. Yet most animals are insects.

The state of play in 1976 was that molecular and cell biologists had essentially struck out when it came to aging, with the tantalizing exception of the work of the cell gang on the Hayflick Limit. Nonetheless, the Hayflick Limit couldn't be a universal regulator of aging. There was no apparent reason to think that aging was going to be a tractable scientific problem in the near future. Yet Brian Charlesworth apparently felt that he had solved the scientific problem of aging, at least theoretically. I was dubious.

What could I say to Brian? I thought that aging was one of the worst possible subjects for a doctoral dissertation. Before I went to Harvard in 1976, I had firmly resolved not to work on aging at the University of Sussex. Such are the resolutions of young graduate students. I came back from Harvard bewildered about what to do for my PhD. Experienced scientists know, however, that such moments of bewilderment often prepare the way for the development of insight. My case was to be no exception.

4

The Force

In the end, one has to make a choice. After my unsatisfactory visit to Harvard, I returned to Canada and Queen's University more open to Brian Charlesworth's siren song from the University of Sussex. One of his points was that I should read some of the earlier work on the evolution of aging. I was happy to do so after the disappointment I had experienced at Harvard.

The literature that Brian directed me to was no older than the 1940s. It had all started with J. B. S. Haldane. Haldane was a figure from the famous Bloomsbury clique that gave the world Bertrand Russell, Aldous and Julian Huxley, Virginia Woolf, John Maynard Keynes, and many lesser figures. J. B. S. Haldane was well known to the Huxleys and the Woolfs; he had played with the Huxley brothers as a child; later Huxley would get ideas for his famous *Brave New World* from Haldane. Aldous Huxley even satirized Haldane as a cuckolded physiologist in his novel *Antic Hay*.

John Maynard Smith had been a student of Haldane's, indeed his greatest student. He once described to me how intimidating Haldane was. "It wasn't just that Prof was very bright," said Maynard Smith. "He was just so much *brighter* than everyone else." Haldane's record of accomplishments speaks for itself. He was a founder of two different scientific fields, biochemistry and genetics, while simultaneously a leading communist intellectual. He was also a great popularizer of science. This suggests to me that Aldous Huxley was getting even with Haldane by satirizing him as a cuckold, for there could have been no doubt as to who had the greater intellect. But I don't suppose that we'll ever know if this psychological hypothesis is in fact correct. There is no question that Haldane had a remarkable talent for offending people; the historical record on this point is quite clear.

The broad swath that Haldane cut through English letters and society left him without time to pursue many of his scientific insights very far. One example of this dereliction was his theoretical analysis of aging and the genetics of Huntington's chorea, as Huntington's disease was then known. This disorder is caused by a single dominant gene. Everyone with a copy of this gene gets the disease, unless they die of some other misfortune first. Huntington's doesn't start to have much effect before 30 years of age in most cases, and often it has little effect until 40. When it starts to kick in, the disease causes nervous system deterioration. Its symptoms begin with a lack of coordination, then move on to impaired intellectual function. Personality is afflicted— victims become impossible to deal with, irascible, and incoherent. Life ends under heavy sedation or confinement, when the nervous system no longer sustains respiration and other basic functions. The entire process of deterioration, from first symptoms to death, can take 10 or 20 years.

The most famous person to suffer from Huntington's disease was Woody Guthrie, the man who most promoted traditional American folk song. (I grew up singing his songs in elementary school classes, particularly "This Land is Your Land.") Woody Guthrie was Bob Dylan's personal idol. Dylan visited him when Guthrie was already living with dementia. The period from first symptoms till death was about two decades. Guthrie had a number of children, one of them Arlo, who carried on the family tradition of folk singing. Because of the eventual devastation meted out to everyone who carries the gene for this disease, the history of the Guthrie family makes for very sad reading. Woody's mother had the gene, and may have been responsible for burning one of her children to death. She had to be committed to a mental hospital, like her son Woody. One of Woody's daughters burned to death as well. His son Arlo, however, is in his late fifties and still performs in public; he is not known to have Huntington's as of this writing.

Haldane's analysis appeared in a book published in 1941, *New Paths in Genetics*. There he proposed that Huntington's disease occurs because the gene creates trouble only after natural selection loses power in hunter-gatherer populations, the kind of populations that made up the entire human species until very recently in our biological evolution. During most of human evolution, on his analysis, people who had the disease gene finished reproduction before the symptoms of the disease took hold, by the age of 30 or 35.

Peter Medawar took up this theme after Haldane. Medawar would eventually win the Nobel Prize for his work on immunology, but before that he produced two papers on aging, "Old Age and Natural

Death" as well as "An Unsolved Problem of Biology," published in 1946 and 1952, respectively. In these works, he proposed that "the force of natural selection weakens with increasing age." This was offered as "the origin of the innate disposition to deterioration with increasing age" (both phrases from 1952). It was the start of significant evolutionary research on the problem of aging.

Unfortunately, Medawar was not an evolutionary geneticist, and he made a muddle of the evolutionary theory of aging at several points. It was a kind of curse—Medawar amplified and named Haldane's concept of the force of natural selection, but then supplied erroneous arguments for it. This may have been one reason why his work had relatively little impact on aging research when it appeared.

A portion of Medawar's thinking was greatly clarified in an article published in 1957 by George C. Williams. I met Williams in the early 1970s, when he visited Queen's University to give a seminar on the mysterious migration of eels in the Atlantic Ocean. Williams is a physically imposing figure, tall, with a Mennonite beard. He speaks slowly, with long pauses. His classic American voice is so even toned that everything he says has the ring of unquestionable authority.

Williams began his 1957 analysis of the evolution of aging by pointing out that biological aging should be distinguished from the breakdown of machines. Animals are not machines but dynamic physiologies, continually taking in nutrition from the environment with which to repair or replace tissues. Even the aging of teeth, which might seem comparable to the wearing out of a car tire, isn't simple wear and tear. In some species, adult teeth continue to grow, as they do in rodents, or they can be replaced, as they are repeatedly in elephants.

Williams also made the important point that animals undergo growth and development in order to produce young adults in excellent health, yet these same adults fall progressively into decrepitude when kept under good conditions in laboratories or suburbs. Surely maintaining vigor should be easier than producing a young adult? If an intact adult can develop from a fertilized egg, why is it so difficult to maintain the vigor of a fully grown adult? Aging is not a reasonable feature of biology. It seems to be a paradox.

Williams explained aging in terms of genes that have multiple effects. This is a common feature of genetics. It happens whenever a single genetic difference produces two or more changes to an organism, such as one gene difference producing both a larger nose and red hair, to give a fanciful example. Williams explained aging using genes that have effects both early and late in life. To illustrate his theory, Williams suggested that a gene that increased the calcification of bone during growth might later cause calcification of the arteries, and thus atherosclerosis. Stronger bones would come in handy for a robustly athletic young person, even though the accumulation of extra calcium would play havoc with the walls of blood vessels later in life. But selection for strong bones in the young would occur anyway, because the vigor of the young is favored by natural selection. So selection for strong bones might cause later cardiovascular disease. In general, natural selection for early benefits might have costs late in life, he thought. This idea was present in Medawar's writing, but he wasn't as clear as Williams.

The next important publication on the evolution of aging was a 1966 paper by William Hamilton, "The Moulding of Senescence by Natural Selection." (In evolutionary biology, the term *senescence* is a synonym for "aging." In botany, it isn't. On the green side of biology, senescence instead refers to the deterioration and eventual loss of flowers and, in deciduous species, leaves.) Hamilton was one of the acknowledged geniuses of evolutionary biology, winner of many scientific prizes, though not a Nobel laureate. No evolutionary biologist has ever won a Nobel Prize, despite many contributions to Physiology or Medicine, the official name of the "biology" category. (Though the Physiology or Medicine Prize has frequently been appropriated by cell and molecular biology, those fields are not part of its definition.) While I didn't have a good understanding of its arguments when I first read it, Hamilton's paper was pregnant with significance. He had graphs and equations that seemed to explain aging in terms of evolution; his curves were in all the right places, yet they were derived from first principles of population biology. I was not yet able to re-derive Hamilton's equations or generate his graphs. That would come later. But what I guessed from the paper was enough to persuade me to accept Brian Charlesworth's offer and turn down Harvard.

I didn't know exactly what I was doing, yet my intuition told me that natural selection was indeed, in some way as yet unknown to me, the key to aging. The problem was perhaps not as inscrutable as I had supposed. I decided to go to the University of Sussex to study aging for my PhD.

5

Goon Show Einstein

It was September of 1976 when I arrived in England. I had just married Frances Wilson, and coming to England was supposed to be a long honeymoon for us. We were typical English-speaking Canadians in our rampant anglophilia, having been brought up to believe that the British Isles were the seat of all that was great in the world's culture, starting with Shakespeare. The University of Sussex housing office had already given us Inner Court Flat in Lancaster House, a name that appealed to our notions of cultural depth. We had visions of a stone building from the sixteenth century and tall Gothic windows.

But the Britain that we found was nothing like our dreams of a green and pleasant land. The trade unions were wrestling with Parliament for control of the country. The traditionally benign climate was in the grips of its longest drought in modern times, and the fields of southern England were brown and yellow. Punk music had just been born; teenagers on the campus of the University of Sussex had green and purple spiked hair; hippies had become utterly unfashionable. Excitement for young people was the Sex Pistols, sniffing glue, and body piercing—things that my wife and I did not understand. Now the clichés of disaffected youth everywhere, the rudimentary music and the self-mutilation amazed me when I first arrived from Canada.

The University of Sussex is a modern "green-fields" British university that opened its doors in the 1960s. A major feature in the founding of the University of Sussex was the abundance of Marxists, indeed communists, who found shelter there. In the 1970s, this gave Sussex a unique climate of egalitarianism and openness. It also attracted the kind of student who drinks a great deal of beer while discussing the latest publications of the Frankfurt School of post-Freudian, post-Marxist, poststructuralist "critical" theory. Their strictly Leninist professors found them baffling. I did too.

The striking thing about the campus of the University of Sussex is the clash between its rolling green hills and the modernist—or are they "postmodern?"—buildings. On one hand, there is tranquil nature—woods, cow pastures, country paths, all reminiscent of Wordsworth on a walk through an ideal English landscape. On the other hand, there is harsh artificiality—the bleak geometry of round windowless buildings with the proportions of dockside holding tanks. Purposeless concrete pillars evoke giant guillotines and gallows. There are even moats without walls that serve no function unless it is the accidental drowning of uncoordinated professors. If there is anywhere that the modern and postmodern have been triumphant, it is on the campuses of universities founded in the 1960s.

Our on-campus apartment turned out to be a ground-floor flat in a modern building erected on the cheap. It was basically one room with a lot of mold and one usable window that looked onto the inner quad of the building. The sky could be seen if you twisted your neck sideways. We would live there for almost three years.

John Maynard Smith founded the Sussex School of Biological Sciences. There he had gathered a coterie of excellent scientists, Paul Harvey and Brian Charlesworth among them. When I appeared, just before the start of the 1976–77 academic year, Maynard Smith was in Michigan, writing his book *The Evolution of Sex*. Brian Charlesworth was away on vacation. It was left to Paul Harvey to show me the ropes, which he did quite kindly. He was the real trendmeister in the group. Youngish, then with long hair and tall boots, he was irresistibly appealing to graduate students. Paul's main research interest has been something called the "comparative method," an approach that he did much to convert from the ad hoc inference of evolutionary patterns to a proper quantitative methodology. He has trained a vast number of doctoral students.

From the rumors that greeted me at Sussex, it was apparent that Brian, by contrast, had had his struggles with graduate students. I was told that one of his graduate students had been taken away in an ambulance for symptoms of psychosis. Another underwent prosecution for some type of fraud to do with his financial support from the university. Yet another had exchanged accusations of academic misconduct with Brian. I never learned the full details of these stories, and I never wanted to. However just or unjust the rumors were, Brian simply terrified the graduate students at the University of Sussex. This was not helped by the fact that he was a sought-after external examiner of doctoral candidates at other universities. In Britain at that time, it was possible for a student to do all the work required for a doctorate, including the preparation of a thesis, but be failed at the last minute in an exam carried out by a single "external" examiner, from another university, and a single "internal" examiner, from the student's own university. The general impression at Sussex was that Brian failed a lot of students when he served as examiner. This did not make him popular with the young people whose lives he could ruin in an afternoon.

Could I have chosen a tougher advisor? I didn't know any better. I came to Sussex as a bumpkin from Canada, unfamiliar with the folkways of English academia. I couldn't manage the false modesty with which British graduate students and faculty alike assiduously burnished their standing. A self-deprecating comment would be followed by the slyest allusion to some recent triumph. My flat Canadian accent couldn't attain upper-class English musicality, with its endless vowels like notes on a French horn. Even the pseudo-working-class dialect favored by the followers of inverted snobbery—educated at Eton but pretending to be from Brixton—was beyond me. I would never sound the part of English graduate student.

The daily round of English institutional life was also a puzzle for me. To start with, most of the day seemed to be taken up with the consumption of fluids: tea, coffee, or beer. There was a half-hour coffee break on arrival at 10:00 AM. At noon there were 90 minutes for a rather liquid pub lunch. Another half hour or so for tea at 3:00 or 4:00. By the time they left at 5:00, the academic staff had put in perhaps four hours of work. How did they get anything done? It wasn't my problem. I didn't drink tea, coffee, or alcohol during my student years. I went to the communal libations only when I needed to find someone, because they were most likely to be in the tea room.

My focus was my work, and that meant Brian. When he returned from his vacation and I finally met him, some weeks into my stay, I was surprised. I had imagined him as an austere thin man, dressed very formally. Instead, he was short, rather bald, with wild hair sprouting from the sides of his head. His features were handsome, but his entire demeanor was that of a comedian. When he wasn't talking about a highly technical point, he wasn't more than a few minutes between jokes. But these jokes weren't humorous anecdotes. They tended to be absurd non sequiturs.

I immediately saw stylistic affinities to Monty Python, and there was some physical resemblance to Terry Jones of the Pythons. But Brian's humor was more akin to that of the Goon Show, the radio program created by Peter Sellers, Spike Milligan, and Harry Secombe in the 1950s, both incredibly dry and wackily surreal. He grew up listening to the show. But whatever the cause, Brian is the funniest scientist I have met in more than 30 years of academia.

Brian was also unusual in his lack of vanity. He was already far-sighted in 1976, and he openly referred to this as a symptom of his declining physical vigor, despite being in his early thirties. He made no attempt to hide his baldness—no comb-over, no toupee. Indeed, his favorite comedy routine was to play the part of an incredibly senile professor, barely aware of his surroundings.

But if Brian Charlesworth was funny, he was also sharper than a Benihana knife. Once he came back from his vacation, he organized a little discussion group at Sussex. I was allowed to attend, despite being the only one in the group without a PhD. Our topic was new developments in evolutionary theory. For me, the amazing thing about our discussions was that Brian could speak in equations, effortlessly rendering them verbally, taking them apart in midair, comparing them with other equations term by term. In 30 years of academia, I have never heard another biologist do that. Mathematicians, certainly. I'm sure J. B. S. Haldane did it, from the stories about him, but he was long dead by 1976.

The peculiar talent that Brian had was the ability to see quantitatively the twists and turns by which evolution works. This is different from the ability to grind out mathematical theory, which many scientists have. We see a problem, we write down some equations, we analyze the equations, and then we find our conclusions. That was how I did theory. Brian could do something more. He could see where a mathematical analysis was going to go long before he even started.

In physics, the greatest theoreticians have this ability. Isaac Newton without doubt had tremendous quantitative intuition. The figure who has been best known for this talent in recent times is Albert Einstein. Einstein saw so much more than his fellow physicists, and he saw it far ahead of the explicitly worked out results of mathematical analysis. That was why he needed technically superlative mathematician colleagues like Hermann Minkowski. Einstein needed them to show formally, by the rules, what he already saw with his intuition and some roughshod mathematics.

Unfortunately I was no Minkowski, so I had to struggle to understand Brian's theoretical analysis of aging. I needed to see the theory worked out in detail, but Brian's papers were very hard to understand. I felt a little better about this when I found him puzzling over one of his own mathematical publications in the lab one day. He told me that he had received a letter from someone who couldn't understand a particular step in his analysis. Reading it over again himself, Brian couldn't follow it either. But it didn't matter. Brian was perfectly capable of finding a different analytical approach that worked even better.

As I tried to find my way to a solid understanding of the evolution of aging, I returned to the 1966 paper that William Hamilton had written. My hope was to develop enough insight from this paper so that I could understand Brian's work, which was more advanced than Hamilton's. But Hamilton's paper contained major mistakes of mathematical notation. Two equations that were printed as ordinary differentiation should have been partial differentiation. That didn't help. I worked with the paper nonetheless. I tried to repeat Hamilton's derivation of the key equations, but I made a mess of it at first. However, over time, Hamilton's paper gave me the leg up that I needed to get over the wall of theory that I faced. I began to understand how aging evolves. The most important thing was to understand the force of natural selection, and explaining it was Hamilton's greatest accomplishment in his 1966 publication.

That academic year I had an opportunity to supplement my reading of Hamilton with attendance at one of his seminars at the University of London. One of my classmates, Michael Orlove, had been a protégé of Hamilton's. Michael is an American, extremely verbal and legally blind, then resembling the hirsute Elliot Gould from the film *M*A*S*H*. We were both complete oddballs relative to the English graduate students, so we were spending a lot of time together. As a Canadian, I guess I was the Donald Sutherland of the duo. Michael needed my help to get to London to hear Hamilton, due to his handicap, and I was happy to oblige.

In person, William Hamilton was a study in dilapidation. His skin seemed almost gray, and his head was enormous. With his bad posture and quietly mumbled words, he was like a horror-movie version of an English butler. His seminar was almost impossible to follow, but he did show remarkable slides of fighting insects, the subject of his talk. I can't say that I got much out of it, but meeting him for the first time was a schoolboy thrill.

On the way back to Victoria Station we were running late, in danger of missing the last train to Brighton that would allow us to make the connection to the Sussex campus. I knew the departure time, but not the platform. Michael Orlove thought he knew the right platform, and we rushed into the station and onto a train that was about to leave. When our tickets were checked, we found that we were on the wrong train, going to Kent instead of Sussex. The ticket collector didn't seem to mind, but we had to double back to London, spending most of the night on British Rail. Michael and I stayed up by talking science, from incest in fig wasps to the evolution of sex. Unfortunately, this offended a middle-aged woman on the train, who screamed at us to shut up. Michael Orlove and I had a unique chemistry that brought out the worst in people.

Now I had met Hamilton the man and I thought I understood his paper on aging after hours working on it. In the same spirit as Haldane's analysis of Huntington's disease, Hamilton's mathematical treatment of aging can be explained using examples from medical genetics. Before reproduction starts in a population, the *force of natural selection* acting on survival is at its maximum. Consider a genetic disease, Hutchinson-Guilford's progeria. The term *progeria* means early aging. This particular form of progeria is also known as "childhood" progeria. It is known in only a few dozen patients at any one time, in the entire world. Progeric children have stunted growth, failure of sexual maturation, and very early cardiovascular disease. Over 80 percent die of heart failure in the second decade of their lives. This form of progeria kills every child who bears even one copy of the disease gene.

Natural selection acts powerfully to keep childhood progeria rare, because the gene for the disorder completely prevents reproduction. This illustrates the general point that the force of natural selection is at its highest level during childhood, because it obliterates dominant genes that kill children, except for new mutations. Natural selection shows no mercy toward the progeria gene, just as the progeria gene shows no mercy to the children that have the gene.

At the other end of life, natural selection snoozes off. One way to understand this abdication by natural selection is to see that the force of natural selection is all about future reproduction, that twinkle in the eye over dinner and a fine wine. If there can't be any reproduction in the future, the force of natural selection is completely deflated. This is

33

the key. If there is no reproduction in your future, natural selection doesn't care about your survival anymore.

Juveniles have their entire reproductive future ahead of them, so natural selection acts powerfully to keep them alive. The force is strong. Thus teenagers are full of vigor, because it is built into their bodies by natural selection. Evolution wants them to stay up late making mischief with their friends. The only solution to this problem with teenagers is to act like a prison warden, which most parents are no longer willing to do.

The postreproductive have no reproductive future, by definition, and as a result natural selection does nothing to ensure their survival or their vigor. We middle-aged slog through our days. The force of natural selection is weak at our age. So we lack strength, speed, virility, fertility, and endurance.

Natural selection is the informational signal of evolution. Working over many generations, it enables fish to get oxygen from water and birds to fly. The force of natural selection reflects the strength of that signal. When we are young, the strong signal supplied by natural selection gives us remarkable vitality. As we get older, the signal fades out. It is like driving out of Los Angeles listening to an LA radio station. (Here I assume that you don't have a satellite digital radio receiver.) The farther you go, the weaker the signal at that frequency. Eventually there is just static. As we get older, we are driving away from the signal of natural selection, and we end up enfeebled as a result.

For evolutionary biologists, the force of natural selection is a set of equations describing the mechanics of evolution at different ages, just as Lorentz transformations are terms in the equations of Einstein's theory of special relativity. The force of natural selection comes from mathematical theory, not metaphor. My task in coming to terms with Brian's thinking was to learn the mathematics of the force of natural selection.

Once I had done this, I was ready to start experimental work. I went to see Brian in the gloomy room that doubled as his office and his laboratory. Brian didn't like to use artificial light. He had been working at his dissecting microscope when I entered wearing a damp trench coat.

"About the aging project," I said. "The force of natural selection must be one of the most unfairly neglected ideas in biology."

Brian put his glasses on and leaned back in his chair with a mysterious expression. Then he got up, walked over toward the gray windows, and quietly laid out in detail the experiments that he wanted me to do. He had snared his reluctant Canadian.

The evolutionary theory of aging had at that point been exposed to little in the way of experimental tests, and none of those tests had been particularly probing. Brian knew that he had to face the question: would his elegant theories survive an experimental challenge?

The research he proposed was based on the fall in the force of natural selection. As the force falls with age, Brian predicted, genetic variation should explode. Let me explain. Consider the impact of genetic mutations. Mutations that affect different ages will experience different forces of natural selection. When natural selection is powerful it throws bad mutations, such as those that cause progeria, out of the population. In this respect, natural selection is the bouncer at the evolutionary nightclub used by teenagers, getting rid of the badly behaved. But when natural selection is weak, at late ages, natural selection acts as if mutations that debilitate older individuals really aren't so bad after all. Natural selection becomes more like a nursing home orderly waiting indifferently for his elderly charges to die. This isn't because these mutations are any less harmful for the health of the old. It's because natural selection has become an underachiever, like Woody Allen's God in the film *Love and Death*. It doesn't really care about the elderly.

Brian's take on this dereliction was quantitative. He doesn't have much taste for metaphors. If we imagine a steady input of mutations with bad effects at distinct ages, mutations that have bad effects at early ages would be stringently eliminated courtesy of natural selection. But when mutations have deleterious effects only at later ages, the weakness of the force of natural selection later in life will allow these mutations to hang around in the population. This makes later life a *garbage can* full of bad genetic information that rots our bodies. What Brian saw was that this evolutionary pattern would allow the preservation of more kinds of genes with bad effects on later ages, but fewer genes with effects on the young. That is to say, there should be more genetic variation later in life in groups of aging animals. If youth is a predictable pattern of smooth skin and ready muscle, the aged should experience a gamut of infirmities, some dying of cancer, others of heart disease, still others of stroke. Youth should be monotonously perfect because of fine-tuned genetics. The old should be diversely decrepit because of out-of-control genetic diversity. This was the prediction that I had to test, the increase in genetic variation with age.

Brian had already decided on the best organism for the experiment: the laboratory fruit fly, *Drosophila melanogaster*. These flies have yellowish bodies, and are considerably smaller than house flies. They are also a lot cleaner. Brian had obtained a large sample of fruit flies from Philip Ives, a scientist who had worked on *Drosophila* population genetics in the United States since the 1930s. (I met Ives in the early 1980s in Amherst, Massachusetts, not far from the place where he had collected the flies Brian and I have used for 30 years. He was as charming as apple pie with vanilla ice cream.)

Brian wanted me to study the number of eggs that females laid in a day, an attribute known to biologists as "fecundity." Fecundity was to be my measure of biological health. Fecundity determines a female's reproductive success, and so her fitness. Following Brian, I expected to find little genetic variation for egg laying at young ages, because natural selection would screen out mutations that decreased egg laying at early ages. But later ages should suffer from weakness in the force of natural selection. Egg laying by older females should therefore be affected by a wide variety of bad genes loitering in the population. The genetic variability of fecundity should increase with age, according to Brian's analysis.

With the specific scientific prediction worked out, the next step was to design the experiment. Scientists spend a lot of time designing experiments, because the fewer surprises you get during the execution of your experiments the more likely it is that you will collect usable data, and thus get your degree, get your papers published, or get a job.

It didn't take me long to realize that if the experiment were to be done correctly, it would involve thousands of flies, data collected over many weeks, and about a million eggs. A single human being couldn't do this by himself. This was a daunting prospect. My solution was to do scaled-down versions of the experiment over and over again, accumulating data as I went. At the end, I planned, I would assemble all the data that I had collected, in one huge analysis. I figured that I would have to do the experiment seven times, collecting fecundity data every day for more than 400 days without interruption. At the end of that, I would have the data to show whether or not genetic variability for egg laying increased with age, as Brian expected.

I started the experiment in the spring of 1977. By the time I finished late in the summer of 1978, I had counted close to a million eggs. Some days I worked in the laboratory for 24 hours straight. The entire experiment took about 3,000 hours to complete.

Shortly after I began this huge experiment, Brian abandoned me to go to North Carolina for a sabbatical. He was planning to write a book there. He came to the lab on a weekend especially to say goodbye, since he knew I would be there. I was always there. For once he didn't crack any jokes, except perhaps when he said, "Don't do anything I would." He almost seemed serious. As it turned out, he had good reason to worry.

When Brian got back from North Carolina, I punched the experimental data onto cards and started the process of reading them photoelectrically, in my effort to persuade the campus mainframe to cough up some results. This was the Stone Age of computing. To get serious computing done, you had to carry around a large box full of punched cards, each card a line of programming or data. To run a program a single time was a cumbersome exercise, and it would take several hours, if not a whole day, to get your output.

One day, I was denied entry to the computing center by a group of campus radicals. They were trying to close down the university for some obscure ideological reason. I never learned what that reason was, but I'm sure it was of the greatest importance. In addition to barricading the entrance to the university, the radicals also wanted to trash the campus computer. This was back before ubiquitous personal computers. University students had access to one or two large mainframe computers, and little else. Knocking out the main computer was indeed a revolutionary plan. It would ruin the lives of untold graduate students like myself. But they hadn't quite broken in to the room that kept the computer. At least not yet.

As I looked on with a sense of helplessness, the ersatz revolutionaries broke through the second to last door to the computer. Then it was on to the last door. Dozens of them pushed together, as I stood there cradling my box of punched cards protectively. But try as they might, they never got through to the computer itself, even though the last door bulged mightily.

And then for some reason, perhaps the coming of spring, the formerly agitated students decided to write their final exams and collect their degrees. They melted away to graduate school, law school, the BBC. This left me free to use the campus computer to calculate what happened to genetic variation in my flies. Brian had predicted that genetic variation would explode with age, as the force of natural selection faded out. And I expected that result too.

But the results were nothing like our expectations. The genetic variation for fecundity later in life was not greater than the variation at

37

earlier ages. Natural selection did not seem to lose its ability to throw out genes with bad effects at later ages. It looked as if Brian was wrong.

This was the first experiment of its kind that anyone had ever performed on any animal. There was no point of comparison, no similar experiment. We were on our own with the results, a rare situation in science. We thought we knew what would happen during the experiment, because Brian's analysis was very persuasive. Apparently, we had missed something. If Brian was upset, he didn't show it. He viewed the situation objectively. If the results said he was wrong, then he could live with that.

The result I got was quite clear. It was a refutation of the garbage-can theory of aging. It suggested that whatever else was going on with aging, the later part of adult life was not just a receptacle in which bad mutations accumulated. Two years of work had left me with a result that undermined every expectation that Brian and I had about aging. Thousands of hours had come to nothing—or rather they had come to a devastating rejection of Brian's idea for my PhD.

If I hadn't misbehaved while Brian was in North Carolina, I don't know that my PhD would have amounted to much. But even the lives of scientists can be sent spinning out of control, and that is what happened to me.

6

Tiny Methuselahs

One of the best of all classic American film noir movies is *The Postman Always Rings Twice*, from 1946. The lead actors, Lana Turner and John Garfield, were not particularly good, though Lana Turner's costumes were incandescently white and her sensuality was laid out for all to see. The great ingredient was the screenplay. Crucial to the film is a pair of fatal car accidents; the film could be seen as 90 minutes of propaganda for wearing seatbelts. At the end of the movie there is a remarkable scene in which John Garfield likens the misfortune of those two car accidents to a postman who always rang twice.

The postman rang for me the first time in the fall of 1977. I was in the middle of my 3,000-hour experiment when my parents came to visit me at Sussex. I had hardly seen them since my wedding, just before my departure for England in 1976. They stopped off to see my wife and I before proceeding to London. Their plan was to go on to Portugal for some sun, after London. The visit went well, Frances having found them a lovely room at an old hotel in Lewes, the historic town near the university campus. One of Henry VIII's wives, Anne of Cleves, had lived in Lewes, and her house still stood on one of the main streets. My parents loved it.

The night after they left, Frances and I were woken by a loud knocking on our door. I checked the time—it was the middle of the night, perhaps 3:00 AM. When I opened the door, my mother burst in, distraught. My brother Tim had killed himself. He had hanged himself from a tree in some woods in Ontario. A close family friend, Major Gross, identified the body. My father was in control of himself, but I could tell that he was shaken to his foundations. Frances held my mother while they cried.

At first, in the days that followed, I was quietly stunned. I did not tell anyone at the university about my brother's death and I didn't go to the funeral in Canada. There was no real plan of action to this. I was simply paralyzed emotionally. I went on with my experiment, working with my flies every single day. They had become a refuge.

I couldn't even think about my brother. We had been too close to be close; less than two years apart in age, we had grown up together as my parents traveled from one military posting to another. In the same high school for a year, for a time we had many friends in common and shared records, clothes, and guitars. We were sometimes mistaken for each other, at a distance. What I couldn't keep out of my mind, after his death, was the time we had gone hiking on the melting ice of a river with a group of friends. We were just boys, maybe I was eight years old and he was seven. There were no adults around. He fell through a hole that suddenly opened up in the ice beneath him. Just as he was disappearing from view, his head beneath the water, we grabbed his hands from either side and pulled him out. From that day on I thought that he was destined to die young, though I never spoke a word about the incident or about my fear.

After the age of 16, my brother's behavior became erratic. Always emotional as a child, as a teenager he lived his life with a charismatic intensity. He was one of those people who could easily engender sympathy, with a combustible mixture of self-confidence and vulnerability. The signs of trouble were not long in coming. He was arrested as a minor while drunk and vandalizing. He spent a lot of money and went through a series of intense relationships with some very pretty girls. He also had periods of acute depression. One day our younger brother Chris found Tim collapsed on the floor of the living room from an overdose of pills. His stomach was pumped. Later Tim injected himself with paint thinner, destroying the tissues of one arm. Physicians set to work to repair the arm. His last suicide attempt, hanging himself from a tree, was definitive. He was not crying out for help; he was determined to end his life, only 20 years of age. It was a devastating blow to all of us, though I was sheltered by distance, and by my research, from my family's trauma. But I had quietly lost any sense of peace.

Then a remarkable thing happened. I began to have new ideas for scientific projects and my reading of the scientific literature accelerated. I was writing down notes at all hours. I bought a series of small blue notebooks so that I wouldn't have to keep track of innumerable small pieces of paper, and I carried the notebooks around everywhere, filling them up with the scrawl that I use for handwriting. One of the pursuits that my rebounding energy engendered was reading the entire literature on aging. Some of it was Taoist stuff, especially from Joseph Needham's series *Science and Civilization in China*. But I read the cell gang literature too, as well as a variety of less cogent material. I would take a journal on aging off a library shelf and look up an article at random. Then I would count up the flaws of logic, experimental design, and analysis. I was using evolutionary theory to reformulate the entire field of aging.

One afternoon, a few weeks after my brother's death, I came across a major find. It was an article by J. M. Wattiaux on the effect of parental age on the offspring produced at that age. The study used fruit flies from the genus *Drosophila*. Since I was working with fruit flies myself, it was inevitable that I would cross paths with this paper. Wattiaux found that when he made each new generation using fruit flies that were the offspring of old parents exclusively, generation after generation, the flies showed an *increased* life span.

This was a lightning bolt for me, because increasing life span was then one of the hardest things to do in research on aging. Decreasing life span in an experiment on aging was nothing special. Many things did that: inbreeding, mutation, radiation, overfeeding, abundant re-production. But scientists hadn't discovered reliable ways to prolong life beyond the limits of what the organism could do for itself.

I realized immediately that if Wattiaux had indeed found a way to produce *longer-lived* animals, the prospects for research on aging would be dramatically elevated. First, because the technique might be applied to humans to make us live longer. Second, because tiny Methuselahs might supply clues as to the controls of aging in general. They had to be doing something better, because they lived longer.

The important thing was to understand how Wattiaux had been able to pull off the postponement, or slowing, of senescence. But Wattiaux didn't know why his fruit flies lived longer. Wattiaux thought that longevity had increased because of a nongenetic effect of parents on their offspring. In thinking this way, he was following a longstanding tradition in biological thought.

For example, it has long been assumed that mothers who undergo emotional trauma during pregnancy give birth to deformed or defec-tive children. Francis Bacon's seventeenth-century book on aging states that the circumstances of conception have an effect on the aging of the child, such as conception in extreme passion, with adultery, or after a long period of abstinence. Pregnant women can still be heard to say that they don't want to look at an ugly man, lest their children end up looking ugly. In fact, I've heard that remark quite a few times. Wattiaux was interested in nongenetic effects of this kind. That's why he had performed his experiment. He also felt that nongenetic effects explained his experimental results, though he had no direct evidence for such effects.

To me, Wattiaux's results pointed to something else altogether: the importance of the force of natural selection. Natural selection discards bad genes, genes like those that cause fatal childhood progeria. Bad genes cause these effects by producing inborn errors of metabolism: letting toxins accumulate, impairing brain function, and so on. Many of the diseases that kill infants are the products of such bad genes, Tay-Sachs disease being one notorious example. Its victims don't live to eight years of age, much less twenty. Afflicted children usually die while still toddlers, blind, in chronic pain. Natural selection keeps genes with such devastating early effects rare, because the afflicted individuals die before reproducing. Bad genes destroy themselves when they kill the young. It's automatic Darwinian justice.

But at later ages, the force of natural selection becomes weak. It leaves genes with late bad effects alone, because natural selection has stopped working. Its force has fallen toward zero. Bad genes that only have late effects will not be removed by natural selection. They can accumulate. There is no more automatic Darwinian screening.

What sets the timing of the force of natural selection? The answer is the age at which reproduction first occurs in a population. Before that age, the force is strong. If reproduction is postponed, the force is high longer. In populations that reproduce early, natural selection declines early. By contrast, populations that are old when they reproduce have powerful selection until they start to reproduce. If the force of natural selection is kept strong at later ages by delaying reproduction generation after generation, selection should increase life span and prolong fertility.

Modern women who delay reproduction favor the evolution of postponed aging in their descendants. However, this is no immediate panacea for human aging, because moderately delayed reproduction will take centuries to have much effect on human aging. Furthermore, the unfortunate side-effect for women who delay having children is that they may find that they or their mate are no longer fertile when they try to get pregnant. This is already a commonplace effect of delayed reproduction in industrialized countries.

Even though he hadn't intended them to, Wattiaux's experiments provided evidence in support of the evolutionary theory of aging. You might wonder why anyone else hadn't performed similar work. The problem is that delayed breeding doesn't give quick results. Fruit fly generations take two weeks in most labs. If a fly lab culture is reproduced using parents that are ten weeks old, you only get a little more than five generations a year. If it takes 50 generations to produce a response, an experiment with late breeding might take ten years. Scientists don't like ten-year experiments for pragmatic reasons, especially their rate of publication and their grant funding. But it had taken Wattiaux only a few years to produce longer-lived fruit flies, and my interpretation was that he had done so by inadvertently forcing the hand of evolution.

This showed that fruit fly experiments prolonging the force of natural selection could get results in a few years. This meant that the force of natural selection could be used as a tool to postpone aging genetically. Time would not be a prohibitive problem.

Yet there was more. If a year or two was enough time to create a fruit fly population with somewhat increased longevity, a longer period with a strong force of natural selection might allow the experimental creation of a population of Methuselahs—animals with substantially increased longevity. A custom-built Methuselah organism would reveal the controls on aging, because an organism that died much later would have to have normal mechanisms of aging slowed or forestalled. This would be the single most important breakthrough in the long history of research on aging: the deliberate creation of longer-lived animals.

To understand why this is so important, consider the problems that geriatricians face, compared to most medical specialists. In studying contagious disease, physicians can compare infected patients with patients free of infection. From that comparison, a great deal of information can be gleaned. Men with syphilis can be compared with men free of the disease. Medical geneticists can compare children afflicted with a heredity disease to children who are free of the disease. But in aging medicine, no one is "normal," because no one is free of the disorder. Everyone deteriorates as adults. There are no exceptional patients who do not suffer from aging.

An alternative stratagem is to compare old patients with young patients. But then physicians find innumerable differences in favor of the young: differences in blood pressure, respiratory volume, maximal heartbeat, immune response, elasticity of the skin, kidney function, and so on. Which of these differences are most important for aging? No one knows.

I knew that making Methuselahs was a far better strategy for the study of aging. Within weeks of reading Wattiaux, in December 1977 I started to make Methuselah flies. I just added that experimental work to my big garbage-can experiment, and worked longer hours. I didn't tell Brian at first, because I didn't want him to say no. He was away in America, and wouldn't find out what I was doing until later. He would also have been concerned about my ability to continue with my garbage-can experiment at the same time that I was starting a completely different experiment.

My Methuselah experiment was simplicity itself. At Sussex, our lab *Drosophila* normally lived their whole lives in old-fashioned, one-pint, glass bottles with a cornmeal-sugar concoction, the "food," on the bottom of the bottle. Flies emerge as adults in about ten days from the start of their lives, which they begin as eggs. A few days later these young adults were transferred to bottles with new food on which to lay their eggs, in turn. The egg laying took just a few hours. Then the adults were knocked out using CO_2, dumped out of the bottle, and discarded. The eggs were then allowed to grow up. The entire cycle took two weeks.

I changed this procedure in my Methuselah population. Each generation was forced to take 35 days, five weeks instead of two. As before, 14 days were allowed for development of the adults. The adults were then kept in bottles with fresh food for three weeks. The bottles were changed every few days, any laid eggs or developing larvae discarded with them. Eggs were collected from adults that were at least 35 days old. These eggs were used to start the next generation. This change of procedure forced natural selection to favor the continued survival and reproduction of adults for three additional weeks. The population reproduced at later ages stayed large in numbers and was healthy.

After 12 generations of this late reproduction, about 60 weeks, I set about comparing the Methuselah flies with normal flies. This should have been a simple procedure, but it was the winter of 1978–79, "the winter of discontent." The English labor unions were flexing their muscles, despite the fact that the Labour Party government of James Callaghan was in office, and the Labour Party was then traditionally the lapdog of the unions. The unions turned the electricity off and on at unpredictable times. The trash wasn't picked up. This was bad for my experiments. During the dark afternoons and evenings of the British winter I used a flashlight to do my experiments when they shut off the power. Some of my fellow graduate students did the same, but many simply went home. Somehow, I was able to finish the experiment without losing any flies or any data.

I analyzed the data in the spring of 1979. When I finished, I virtually floated up the stairs to Brian's office to present my findings to him. When he had absorbed the results, Brian leaned back in his chair, rubbed the side of his head vigorously and said, "Well, that's it then." He realized straight away that I finally had something good enough get my PhD.

My results were similar to those of Wattiaux. My Methuselah files lived about 10 percent longer. Prolonging the force of natural selection easily postponed aging. I could make tiny flying Methuselahs at will.

This experiment also showed that the theory developed by Hamilton and Charlesworth was correct. The force of natural selection determined aging. Decades of evolutionary theory were vindicated by this result. Brian and I were satisfied that we had shown experimentally why aging occurred. In a crude sense, we felt that we had demonstrably solved the problem of aging using evolutionary theory. After all my initial skepticism, aging had turned out to be a peach. And I was sure of getting my PhD. Somehow, out of the ashes of losing my brother and the demoralization of my family, I had found a latter-day philosopher's stone. The consequences of this finding have dominated my career ever since, as you will see.

7

The Postman Rings Again

In the first part of 1979, before getting my PhD, I was called to the phone while working in the lab. This was England, still in its dark age, and it was hard to get a phone for oneself. Since I had lived without one at my flat for almost three years, receiving phone calls was a rarity in my life. And the name of the caller was unfamiliar to me.

It was Leo Luckinbill calling from Michigan. I stood there in my lab coat, in somebody else's lab, listening to this stranger ask about the design of my experiments. He had heard what I was doing from my former office-mate at Sussex, Bill Moore, with whom I had discussed my research. Leo wanted to do fruit fly experiments that were similar to mine. We spoke for some minutes, but somehow failed to communicate. Still, we agreed to stay in touch.

In September 1979, Brian and I wrote up my doctoral results in a short paper for the journal *Nature*: "A test of evolutionary theories of senescence." It was rejected straightaway on the grounds that we hadn't explained the data using nongenetic theories like Wattiaux's. I took this rejection quite passively; I knew very well how bad the field of aging research was. I had doubted that Brian's sort of theory would receive much of a welcome from the cell gang, and they were the best gerontologists around.

By contrast, John Maynard Smith was furious about *Nature*'s rejection. It was as if the editors of *Nature* were rejecting evolutionary theories of aging by fiat. As Maynard Smith was one of the world's leading proponents of evolution as a scientific theory, this aroused his ire in the extreme. He complained to the editors of the journal—called them out onto the carpet, really. I never learned the details of the exchange, but I got the impression that the *Nature* staff were rather embarrassed. A new review of our paper took place, and a final version of the article appeared in *Nature* in the autumn of 1980. I was surprised that reason had prevailed, but very pleased.

My doctoral examination took place in John Maynard Smith's office at the end of the summer of 1979. There was a small, framed, B&W photograph of J. B. S. Haldane in attendance. I have a picture of Maynard Smith in his office, and it shows that photo of Haldane just above John's head. Douglas S. Falconer, the great quantitative geneticist, was my external examiner. John was the internal examiner. By the rules of British academia, neither Brian nor anyone else was allowed to attend. The three of us sat in rundown chairs facing each other. There was no table to hide behind, nor was I required to stand. It was the most casual "doctoral defense" I have ever attended.

Falconer was very nice to me during the examination, but Maynard Smith was something else altogether. John challenged me to explain why my research wasn't entirely redundant after his 1950s work on aging and *Drosophila* female sterility. (I take up this work in the next chapter.) I admitted the suggestiveness of his earlier work, but wouldn't renounce the claims that I was making for my research. I think John was toying with me, so that my doctoral exam would be more of an event. But I can't say that for sure.

I left Sussex for Madison, Wisconsin, soon after my PhD was approved at the beginning of autumn, 1979. It was great to have a PhD in my early twenties, but I was more or less exhausted by the thousands of hours of lab work. While I was at Sussex, I had never had time to cross the channel for Paris and the "Continent," just a few miles away. I had taken a few hours off in London from time to time, just 60 miles to the north. But I had taken only one vacation, a week in Bath. I was burned out. The last few months writing my thesis had required 14- to 18-hour days, seven days a week. My wife, Frances, left England early, fed up with my neglect of our marriage. I had lost a lot of weight, and had few pleasures beyond science and my perennial addiction, music. So when I arrived in 1979 for postdoctoral study at the University of Wisconsin, I arranged my experimental life so that I could have whole days off, even weekends. I resumed a normal academic life, after the extremes of my doctoral stint.

At the University of Wisconsin I worked in the lab of James F. Crow, one of the leading population geneticists of the twentieth century. Everyone who knows Jim Crow realizes that he is one of the most genial, thoughtful, and helpful of scientists. Some of us openly refer to him as The Perfect Man, for his remarkable combination of character and intellect. Let's face it, most academics are either arrogantly self-confident or reserved, sometimes both. I know I oscillate between the two. Crow is the singular exception, for reasons that none of us have figured out. He was raised a Quaker, and was a pacifist during WWII, but there are plenty of obnoxious pacifists. In addition to many delightful contributions to the theory of population genetics, Crow was also primarily responsible for bringing Japanese genetics back into the scientific mainstream after WWII, particularly in his role as a mentor and sponsor of Motoo Kimura, perhaps Japan's greatest biologist in the twentieth century.

In addition to the Crow lab, Sewall Wright worked just two floors down in the genetics building. Sewall Wright was unequivocally one of the greatest scientists of the twentieth century, a pioneer in several fields: physiological genetics, evolutionary theory, and multivariate statistics. He turned 90 the year I arrived in Madison, but he still had a massive intelligence. His seminars were long disasters, because he no longer had any time-management skills, but talking with him one-on-one was like reading the most profound book ever written in biology. Wright would speak to you in paragraph after paragraph of perfect sentences, compromising only with respect to the page numbers of the articles he cited in midair. A quiet, self-effacing man, on the reserved end of the academic spectrum, Sewall Wright was nevertheless a god of evolutionary biology.

Some months after I moved to the American Midwest, Leo Luckinbill got in touch again and arranged for me to visit his lab at Wayne State University in July 1980. He picked me up at the Detroit airport with great enthusiasm and a van. I stayed at his house, where his family co-existed with his restless intellectual life and a small incubator that he used for experiments on microbial ecology. Leo was a marvel of American enthusiasm and intellectual rigor. Working closely with Leo was Robert Arking, the older of the two, whom you met in chapter 1. Bob Arking has always reminded me of Al Pacino. It's probably the urban Northeast accent and the droopy eyes. Bob was the one who contributed the fruit fly expertise, while Leo was the one who best saw the connections between experiments and mathematical theory. I thought that they were well matched.

Leo and Bob were trying to study the evolution of aging in *Drosophila*, and they were interested in using genetic techniques and selection. I had been doing exactly that for four years, so I thought that I could help them. During my visit, I talked them out of doing the heartbreaking genetic garbage-can experiment that I had done for Brian

Charlesworth. I saw no reason why anyone else should suffer the way I had, notwithstanding the fact that Leo and Bob were going to throw ten times the manpower at the problem. They had a large lab full of assistants and students, because they were well funded by the National Institute on Aging. At Sussex, in Third World England—as it then was, I had been forced to scrounge for the most meager of supplies. It was my first exposure to the munificence of American science, and the impression stuck.

I also proposed to Leo and Bob that they start their own Methuselah lines, like the one I had created at the University of Sussex. They had already been thinking of doing this experiment, because they too knew Wattiaux's work well. For my part, I was so enamored with the approach that I had already started new fly Methuselah lines in February 1980, but on a much larger scale. It would improve the scientific standing of the force of natural selection if multiple laboratories used the force to make tiny Methuselahs. Leo and Bob would be using different fruit flies, different fly food, and different incubators from mine, but the scientific designs would be parallel. Leo and Bob, at that time, were some of the most open and congenial scientists I had ever met. It didn't take them long to agree. I helped them with a few tricks of experimental design, and by the fall of 1980 they too were breeding Methuselah flies.

The fall of 1980 also marked a crisis in my personal life. My marriage to Frances had foundered badly. I worked too much, while she hardly had a day of paid work since our marriage, not that she sought employment in any event. I was sometimes poor company, while she alternated between extremes of depression and energy.

Two events had driven Frances far of course: the suicide of my brother and the murder of her stepbrother Eric. I have already mentioned how we came to learn of Tim's death. Frances and Tim had enjoyed an excellent rapport, though they spent little time together. Frances had also been quite close to Eric. When he had been in his early teenaged years and she was in her early twenties, he had fallen under the spell of her physical and emotional magnetism. Eric was knifed to death in the van that Frances and I had taken on our honeymoon in New Brunswick. A film was made of his disappearance and the eventual uncovering of his murder: *Just Another Missing Kid* (1982). The brutality of the story left her badly shaken.

But there was much more to her than periods of depression. She also suffered from delusions and periods of intense physical restlessness. When

49

her energy levels reached feverish intensity, she would leave me for days or weeks at a time. The summer of 1980 marked the start of one of her longer vacations from our marriage, and I doubted whether she would come back to me. I became depressed. I still went to work, but every waking moment was dominated by pain that had no particular location.

When our marriage almost seemed to be over, in October of 1980, we found a new plan for our lives. I would take a faculty position at Dalhousie University, located in the city of Halifax, part of the Canadian province of Nova Scotia. We both missed Canada by that time, and hoped that a return to our native country would revive our fortunes as a couple. We returned to Canada in June 1981.

It was not to be. In the spring of 1982, we legally separated and she left for Kingston, Ontario, where we had first met nine years earlier. I never saw her again, though we remained in touch by phone and vacillated about getting a divorce. I was struggling with depression and solitude through the rest of 1982, but early 1983 saw me rebounding. Frances seemed to be feeling better in Kingston too, though I only spoke to her at intervals, so it might have been a façade.

A sequence of seminars on the West Coast took me to California and Washington State in February of that year. My last stop was Seattle, where I was speaking at the University of Washington. I gave a talk the morning of February 28, 1983; it concerned some of my heterodox ideas about sex and speciation. After lunch, I was supposed to give a talk about aging. My host was Professor Montgomery Slatkin, whom I had met back in Sussex. Monty was going through a divorce in 1983, and he seemed rather uneasy.

Just as we were gathering to go out for lunch, a middle-aged police officer appeared carrying a telegram. The man seemed quite subdued, unusually so for a police officer on duty. Monty froze; he thought something had happened to his wife. But my suspicion was different and, as it turned out, correct.

The police officer asked for me. I came forward, anticipating the blow. He told me that my wife had killed herself. There had been screaming coming from her Kingston apartment all through the night. Her skin had been blue, suggesting cyanide. I spoke out loud, but to myself. "It's finally over."

The police officer obviously regretted giving me the news, and left shortly thereafter. Monty was visibly agitated, looking at my face every few seconds, then looking away. The colleagues who were supposed to have lunch with us were quietly sent away and Monty led me off. I can't remember where we went, or what I ate, but I do remember Monty suddenly unburdening himself with the story of his divorce. I think he had finally found that there was something that he wanted to talk about even less. He had known Frances from our days at Sussex, and like everyone else, he had not failed to notice her. I tried to be a supportive listener, but we both knew I was only going through the motions.

Monty wanted to cancel my afternoon seminar, which was to be given to a fairly large audience. He also proposed dropping the scheduled one-on-one meetings with my other colleagues at the University of Washington. At first I didn't know what to do, but after lunch I decided to go ahead with the afternoon as planned. Later that day I gave the most difficult academic speech of my life; nothing since has even distantly compared.

That evening was the broadcast of the final episode of the *M*A*S*H* television series. I was staying with Barry Sinervo, a student of Monty's, who was giving a party to mark the historic occasion. I don't think that many of the people at the party knew what was up with me, or cared. I drank heavily in order to get to sleep. I would consume more alcohol that year than at any other time in my life.

This time, when the postman rang again, everyone knew. I broke down in tears once during a colleague's seminar, for which I was deeply ashamed. At the Academy Awards, broadcast that year in April, the film about Eric Wilson's murder, *Just Another Missing Kid*, won the Oscar for Best Documentary. The brief clip they put onscreen during the ceremony showed the VW minivan that Eric was killed in. It was the minivan Frances and I had taken on our honeymoon in New Brunswick. That was not an easy night.

But my depression did not last. I began a rapid-fire series of scientific projects and personal relationships. I sprang back to action with a determination that would dominate the next decade of my life. The science of those years has worked out pretty well. My output of publications leapt upward that year, and has yet to fall back in volume. My personal relationships have not fared as well. I have never completely recovered from my first marriage, despite two further attempts, both failures, and four children. My work has been my shelter from life.

One of the immediate benefits of my neurotically expanded productivity was the publication of the paper in which I first presented my relatively famous "Methuselah Flies." (This project is presented in detail in the 2004 book of that name, referenced at the start of the bibliography.) It was submitted to *Evolution* in May of 1983. It finally appeared in 1984 alongside a paper from Leo Luckinbill, Bob Arking, and their students that described the independently executed experiments at Wayne State University. The two labs found essentially the same result: life span could be increased significantly by delaying reproduction over enough generations. The number of generations of selection and the increase in life span published in these papers were even greater than Brian and I reported in our 1980 article in *Nature*. This is an ideal situation in science, to have independent experiments that support a theory. Science absolutely requires independent confirmations of experimental findings.

By 1984, publications from Leo Luckinbill, Bob Arking, Brian Charlesworth, and myself had brought the evolutionary theory of aging to the forefront of evolutionary biology. We had used the force of natural selection to change aging in an insect species, which supported the theory. It was a good start.

It also marked a change in my attitude. Eight years earlier, my interest in aging had been grudging, as I have admitted. Brian had pushed hard to get me to work on the problem. But by 1984, I was dedicated to it. I saw the question of aging, including the problem of increasing the human life span, by the light of evolutionary theory. I was anxious to sort out all aspects of aging in terms of evolution, and so I would proceed.

Death had made me less of a dilettante about aging. I had made its acquaintance in a very personal way, and I did not find it an engaging companion. Let the Christian moralists swoon in its chilly embrace. I find their graveyard desires revolting.

8

Cheshire Cat Cost

There was a telltale clue in the data from the Methuselah flies that I bred at Sussex, a tug on the thread of mortality. As I had expected, the fecundity of Methuselah flies was much greater later in life. The increased force of natural selection in these flies was predicted to increase their later fertility, along with later survival, over the course of their laboratory evolution. That wasn't the clue.

The clue was a dramatic drop in early fecundity in the Methuselah flies once they had evolved increased life span. Their postponed aging evidently came with reduced early fertility. To live longer, early reproduction had to be reined in. This was an echo of George C. Williams, and his emphasis on the idea that natural selection would prefer youth over old age, whenever there was a conflict. It appeared there was a conflict in my flies. Too much youthful reproduction killed, according to the results of the selection experiment.

I went back to the data from my genetic variation experiment, my biggest research effort during the 1970s, but also my biggest frustration. When I looked at the genetic relationship between early reproduction and longevity in these data, the same pattern showed up. Genes that increased early reproduction also decreased life span. This was reassuring, because this finding fit the one obtained from selection, and especially because it came from an experiment with a different design, lacking selection.

In the early 1980s, the Methuselah fly-breeding experiments of Leo Luckinbill, Robert Arking, and myself indicated that there was a cost to early reproduction when longevity increased. All of our longer-lived lines also had reduced early fecundity. There was an intriguing result like this in one of Wattiaux's original studies from the 1960s. His longer-lived fruit flies tended to have reduced male fertility at early ages. Later work on Methuselah flies by my postdoctoral student Phil Service found reduced sexual competitiveness in my young Methuselah males too. In both males and females, early fly reproduction is reduced when selection favors later reproduction.

Using completely different methods, Maynard Smith published evidence for a cost of reproduction in 1958. This was the work that he had confronted me with during my doctoral exam in 1979. In his experiments, female fruit fly reproduction was deliberately reduced three different ways. The first was to deny females sperm. This doesn't completely eliminate egg laying in fruit flies, but it does reduce it. Maynard Smith showed that lifelong virgins live significantly longer than normally mated females.

But since virgins still produce eggs in fruit flies, a better test of the effects of reproduction would be to eliminate egg laying altogether. There is a simple way to curtail fly egg production: sterilization by irradiation. Irradiated females live even longer than virgin females.

Irradiated females still have reproductive structures, most importantly ovaries. In fruit flies, the ovaries are large structures, up to 30 percent of the total body weight. Fruit flies have mutants that yield females that wholly lack ovaries. These *ovariless* females lived longest of all in Maynard Smith's experiments. The physiological burdens of sex and reproduction kill females. Contemporary genetic research on aging in fruit flies, and even mice, has continued to supply additional examples of mutants that reduce female reproduction and give increased life span.

Male fruit flies aren't that different with respect to the cost of reproduction for adult survival. Just because fruit flies are tiny insects, don't think that they have a simple sex life. Male fruit flies are impressively sexy beasts. They have elaborate courtship. In some species the male mouth parts are used to stimulate the female genitalia. Copulation can last much longer than the minute or two characteristic of mammals. It is not unusual for fruit flies to sustain intromission for 20 minutes or more. During this time, the female carries the male around on her back, feeding during the fornication. This is quite the variation on the traditional dinner date.

Fruit fly males will go from one bout of courtship and copulation to another in hours, sometimes in minutes. If the females are willing, which they are almost sure to be if they're virgins, males will mate with six or more females in the course of ten or twelve hours. And they will do it again the next day. And the next. And the next. The only problem is that the males are unlikely to live more than two weeks on this regimen of frequent copulation. Male fruit flies can truly copulate themselves to death. Lotharios take note.

Fruit flies have apparently read the Williams 1957 paper on aging. That is, they have a trade-off between reproduction and survival. Joni Mitchell put it in fewer syllables in her song "Sex Kills."

This finding connects aging research to enduring prejudices and cultural myths. In Western European culture, sex is expected to lead to bad outcomes. For example, the girls in horror films who have sex are the ones who die first. The virgins are usually spared decapitation, bloodsucking, and the rest. At least it seems that way.

Traditional Christianity disapproved of sex. Celibacy was considered a good thing. The official celibacy of Catholic priests is the most obvious manifestation of this Christian sentiment. Actual priestly practice has often been different, but it is the negative view of sex that was official Church doctrine.

Another tradition connecting sex and aging is that of the Taoists of ancient China. A central Taoist practice is the production of *ching*, and its retention within the body. In cosmological terms, *ching* was supposed to be the "life force" or generative essence. Taoists equated *ching* with semen in males and menses in females. In their religious practices, Taoist males used to copulate with a succession of women in a single day, without ejaculation. Ejaculation was either avoided altogether or the ejaculate was diverted up into the urinary bladder by manual pressure on the male urethra. The goal was to stimulate the production of *ching*, identified with semen, but prevent its loss. With more stored *ching*, the Taoists think, the body is better able to withstand the ravages of aging. Obviously this is not a Western point of view.

Fruit fly research supports the view that food energy and materials that are expended in reproduction might otherwise keep the body alive. This is a Taoist kind of result. But does this cost of reproduction apply throughout the living world? With most species, we can't do the experiments that we have performed with fruit flies. To survey the living world impartially, we need experimental procedures that will work with a variety of species.

One such procedure is castration. In biology, the term is much more general than in human affairs. Biological castration means the destruction of the reproductive organs of either sex. It applies equally to the destruction of reproductive tissues in female fish, male horses, and hermaphroditic flowering plants.

It is usually easy to destroy reproductive organs. In some species, they are located outside the body wall, as are mammalian testes, or in distinct structures that are easily removed, as are flowers. Since reproductive organs are almost never required to keep the body alive, they can usually be removed without killing the whole organism.

Sometimes castration is difficult. The ovaries of mammalian females, for example, are relatively inaccessible. But even in such a challenging case, good surgical technique makes the removal of ovaries possible without collateral harm. Whether such an elaborate intervention is feasible in an experimental study is another question. But castration is often possible, however difficult.

Consider the castration of Pacific salmon. Pacific salmon usually reproduce only once, dying immediately afterward. This looks like a well-defined example of the cost of reproduction. But this piece of natural history, as colorful as it is, does not directly support the idea. It might be that reproduction and death are on the same timetable, both triggered by the swim back to the natal freshwater stream for breeding. But we know that isn't it, because castrated salmon can live years longer than intact salmon. Castration of the salmon shows that it was reproduction that was killing the unaltered fish.

Soybean plants show the same effect. Like Pacific salmon, soybean plants only reproduce once. After reproduction, the plant dies. Castration by stripping all the flowers off a soybean plant allows it to go on living for months longer than intact soybean. Even in plants, castration can prolong life.

There is a mammalian species that shows the same pattern as the Pacific salmon and the soybean, the brown Antechinus. This creature is a member of the marsupial group of mammals, together with kangaroos, koalas, and Tasmanian devils. Small and mouselike, it inhabits southern Australia. The brown Antechinus is casually referred to as the marsupial mouse, but it isn't a mouse at all. This animal reproduces according to a yearly cycle. Males live less than a year. There is only one short mating season. During the mating season, males develop enlarged testes and become highly aggressive. They fight, court, and copulate for a period of several weeks, with little or no feeding and less rest. Then they drop dead. Once again, if these males are castrated well before the mating season, they can live months longer. This is a particularly acute case of the rule that sex kills.

With this backdrop, we can move on to human castration. Eunuchs have been known throughout human history, from the castrati of Papal choirs to the harem guards of the Ottoman Empire to the eunuchs of China's Forbidden City. It is frequently convenient for the rich or powerful to mutilate boys. Thanks to this manifest cruelty, there is a reasonable amount of eunuch natural history that is known.

Eunuchs do not develop the lower voice of intact men, but that is not to say that their voices are the same as those of boys or women. The historical record indicates that the Vatican castrati had unique voices, like nothing else. But this may be exaggeration. Eunuchs tend to be taller than intact men. Apparently testosterone inhibits the growth of long bones. Eunuchs can have erections, and some copulate. Castrated men ejaculate a clear seminal fluid. Eunuchs do fall in love. Casanova recounts the story of one of his early mistresses, Teresa, whom he met while she was masquerading as a castrato. She had had a previous relationship with a touring castrato who taught her how to imitate castrati on stage and, more unusually, in the bedroom. (This was eighteenth-century Europe, when clothing was suited to concealment and deception.) The castrato even supplied her with a prosthetic penis, with which Teresa could satisfy female lovers. This device caused Casanova some confusion when he fell in love with Teresa, since his hope had been that she wasn't a man. His reaction to placing his hand on her rather convincing prosthetic is one of the funniest moments in literature.

There are interesting data from twentieth-century American eunuchs. Victims of eugenic legislation, many men with low IQs were castrated in American institutions. Compared with intact males kept in the same institutions, the eunuchs died less often. Their longevities were significantly increased. Even in humans, castration helps survival. When I mention these data in public, however, the verbal response from some men in the audience is that castration just makes life feel longer. On the other hand, I have met quite old castrated men who are proud of their great longevity and their castration.

Castration is an extreme alteration. Can we detect a trade-off between reproduction and survival without castration? In the fruit fly data from Maynard Smith, myself, and others, moderate reductions in fertility also go with increased longevity. These data and the effects of castration together support the general idea of a cost of reproduction.

Are there other data that support the cost principle?

One source of such data is the comparison of species. It has often been found that species that have high rates of reproduction tend to have shorter life spans. This is well known in mammals, where mice epitomize rapid reproduction and short life span while elephants epitomize the other extreme. Humans are much like elephants: we have a long gestation, a slow process of maturation, a low rate of reproduction, and finally a long life span. However, comparing the life spans of biological species can be complicated, which we will talk about in the next chapter.

Over the full range of data, from fruit flies to castrated salmon, the evidence suggests that reproduction curtails lives. Natural selection lets this happen because it doesn't care about the later price, when there is a benefit during youth. We seemed to have discovered a major bit of the machinery by which evolution produces aging.

But one day things changed.

I had been interested in the effects of reduced nutrition on aging. Mice and rats that are given 15 to 40 percent fewer calories live about that much longer. I was curious to see if the same was true of fruit flies.

As good fortune would have it, in the fall of 1989 I was able to recruit two exceptional graduate students: Adam Chippindale and Armand Leroi. Both were Canadian. Adam had grown up in Ottawa, Ontario, but Armand was the son of a diplomat, and had been born and mostly brought up abroad. I taught Armand when he was an undergraduate, and I had a high opinion of him. Adam had been a friend of Armand's, and Armand played a considerable role in persuading Adam to consider applying to work in my lab. To my good fortune, the two of them ended up with me.

As Brian had treated me, I allowed Adam and Armand a few months to settle down before I gave them the nutrition project. For their first experiment, I got them to monitor life span and egg laying, with high and low levels of nutrition.

After some grumbling about their lack of interest in aging, which sounded like my younger self, Adam and Armand set to work. They turned in a very clean experiment. We found that reduced nutrition greatly reduced egg laying, but it extended life span at the same time. This would turn out to be very interesting, but it wasn't surprising. Irrespective of nutrition, the Methuselah flies always lived longer. Nice, but no revelation.

The real shocker was that the longer-lived flies had superior egg laying at every age, compared to normal flies, even as young adults. The cost of reproduction had seemingly disappeared. The longer-lived flies had it all: increased length of life and more reproduction along the way. It was one of the most confusing moments of my research career. Could George C. Williams be wrong about the trade-off between youthful reproduction and survival to old age, after all? If he was, how could we explain the evolution of aging?

I knew that some of my colleagues hadn't always found the trade-off between longevity and reproduction that Maynard Smith, Luckinbill, Arking, and I had. But I tended to think that this disparity was due to badly designed experiments in other labs.

My initial reaction to our own contradictory data was that just such an experimental design problem might be obscuring the trade-off. I remember saying to Adam and Armand, the day that they showed me their anomalous results, that it must be an environmental effect. It was an easy thing to say off the cuff. It took three years to figure out what was really going on.

As a scientist, if you suppose that some mysterious environmental effect is ruining your experiment, you have to find it and reverse it. Just speculating that there might be such an effect is not enough in experimental science. We had to solve this puzzle very explicitly.

One possibility was the environment the flies lived in. The normal flies used in our experiments were cultured in vials that had "spent medium," a polite term for fouled food, food that flies had died in or had excreted in. The longer-lived flies spent most of their adult lives in good conditions, without the rotting bodies of other flies. Spent medium favored the control flies when we tested for its effects; their early fecundity suffered less from such an appalling environment. But the controls were still a bit inferior, providing we gave the flies a lot of time for egg laying. We weren't quite there yet.

The final resolution came when the flies were given spent medium and very short periods of time for egg laying. This was how the normal flies had been maintained for years. Under these conditions, the longer-lived flies were clearly inferior in early fecundity. This was the environment that we should have used in our first nutrition experiment, because it was the normal ancestral environment for all our stocks. The environment that we used in our nutrition experiment was too benign for a trade-off to occur between egg laying and survival, because that environment featured abundant food and lots of time for egg laying. The original trade-off was recovered only by employing the correct environmental conditions. I was tremendously relieved.

Later I would name this kind of difficulty in evolutionary research the Cheshire Cat Syndrome, in honor of the cat that appeared and disappeared in *Alice's Adventures in Wonderland.* Appearing and disappearing costs of reproduction occur sufficiently often that I felt the phenomenon deserved a name that would draw attention to it.

This recovery of a trade-off between reproduction and survival under evolutionarily appropriate conditions restored the credibility of a genetic mechanism that explained the evolution of aging: genetic trade-offs between later ages and early ages. Specifically, early reproduction tends to trade off against later survival. And since the weakening force of natural selection tilts the balance against later survival, evolution will tend to produce vigorously oversexed youth and decrepit old age, because it favors genes that enhance early reproduction at the expense of later survival. Yet this can be hard to detect experimentally. However, it should be understood that this is *not* the only genetic mechanism by which aging evolves. It is merely one possibility for the evolutionary genetics of aging, a scenario that we are sure does arise in at least some cases.

An interesting historical note is provided by the discovery of similar findings in a recent (2003) article concerning trade-offs in the longer-lived fruit fly mutant *Indy.* Kept under good conditions, this mutant enjoys a doubling of average adult life span, an increase in total fecundity, normal metabolic rates, and unaltered flight velocity, all findings that are comparable to results with the studies of our longer-lived flies produced by laboratory evolution, as outlined in our *Methuselah Flies* book and the many previously published studies that it includes. But when *Indy* flies were given lower-calorie food, they were inferior to "wild-type" flies with respect to reproduction. In other words, the trade-off was recovered under stressful conditions, exactly as we found. However, because this was a genetic study with randomly generated genotypes and arbitrary environments, there is no straightforward interpretation of this finding, unlike our earlier work where evolutionary theory supplied such an interpretation. The results from my lab show that populations reproduced at early ages under bad conditions for many generations are likewise more effective at reproducing at early ages when conditions become difficult. It is always entertaining when molecular biologists rediscover findings from evolutionary biology. They have such an appealing naïveté, like the moment when my son Darius gleefully discovered at the age of one that gravity would help him knock over a glass of milk.

Why is it sometimes difficult to detect trade-offs between survival and reproduction? Fly Methuselahs robustly show increased longevity. Even with large changes in nutrition and handling, we still see increased longevity in Methuselah flies. *Life-span differences* between longer-lived flies and their controls don't show much sensitivity to the environment. The longer-lived flies are almost always superior. Obviously this wouldn't be true in some extremely destructive environments, but it is true even in such deadly environments as those that lack all food or that lack any humidity. Fly Methuselahs are just generally better at surviving, even if not universally so.

But from lab to lab, and even within labs as we have just seen, the connection between longevity and reproduction has been unpredictable. Reproductive trade-offs with aging come and go. The problem lies in the distinctly different relationships that early reproduction and later survival have with natural selection. Natural selection acts powerfully on early reproduction. Therefore, evolution will make organisms respond to their environments so as to maximize reproductive output early in life. If there is more food, then sex, reproduction, and death should come earlier. If there is less food, the whole life history should be stretched out. Early reproduction will be highly responsive to the environment.

Meanwhile physiological mechanisms specific to later survival are left unshaped by natural selection, because selection has stopped caring about the continued survival of the old organism. If there are potential physiological responses to environmental change that might enhance later survival, natural selection on its own will not do a good job of evolving these responses. There won't be as much finely tuned sensitivity to the environment. Old codgers are left heedless, whether they are fruit flies or retired plumbers. There is an inert quality to the later part of life.

It is hard for scientists to design experiments that reveal a cost to reproduction. Even if there is an antagonism between reproduction and survival, the exquisite sensitivity of reproduction to environmental conditions will tend to obscure that antagonism. This reconciles failures to find trade-offs with the experiments that have clearly shown that these trade-offs exist. Some things are hard to do in science, and showing that sex kills is one of those hard things.

This also explains the fact that drastic procedures like castration produce increased life span more reliably. Castration physically removes the reproductive system whose responsiveness is so confusing when it is intact. Castration is often the most revealing experiment that can be performed on the relationship between reproduction and aging. Genetically obliterating reproductive structures is even more revealing than surgical castration, in principle, but there are very few organisms in which we know how to achieve such obliteration.

As with animal studies, the study of human castration is only a crude indication of the relationship between survival and reproduction. A problem affecting all studies of human reproduction and aging is that humans live a long time. This makes it hard to collect good data. It is also hard to get volunteers for human castration experiments.

It is doubtful that we will ever have perfect evidence concerning the cost of reproduction in humans. For now, the available evidence supports the idea that humans are like flies, soybean, and salmon in having a trade-off between reproduction and survival.

What are the implications of this trade-off for your life span? It may indeed be better to be a wall flower. That is, the costs of more active reproduction may reduce human life span. It is important to understand that the costs of reproduction start with the growth of reproductive organs and genitals, then proceed through competition for mates, courtship, and sex, reaching a high point with pregnancy and care of the human neonate. Sometimes the costs of reproduction continue through the offspring's graduate studies. On the other hand, these are some of the most important of all human activities, and can be the most rewarding. Like migrating salmon or intact marsupial mice, we may be unable to resist our natural inclinations. Reproduction is one of the enduring human goals, yet it may bring with it our downfall.

9

Birds and Bees

Ford Doolittle is a colleague of mine at Dalhousie University, a molecular biologist of powerful imagination. But that's not why I mention Ford here. After his fortieth birthday, Ford was given to ruminating about his mortality. It particularly bothered Ford that his pet bird, a member of the parrot group, might outlive him. He would stare at the bird in its cage with a mixture of affection and envy. He asked me, why is it that animals from some species live a long time, but others die off early? It is a question that Aristotle asked millennia ago. This is the question that I address in this chapter.

There are many beliefs about how long animal and plant species live. Most of them aren't true. There are no magic birds that live forever. But patterns of aging, among species, among breeds, and among sexes reveal a great deal about aging.

Ford was right about parrots. Reliable reports of parrots living over 40 years are common. Cockatoos can live up to 60 years, and there are some claims that they live as long as 120 years. Ford might indeed be outlived by a young parrot.

Outside of the Doolittle household, the interesting contrast is not between Ford and his parrot. It is the contrast between birds and mammals, as groups. Both mammals and birds maintain a stable, high, body temperature, unlike the vast majority of animal species. Therefore their metabolisms are broadly similar. Yet birds usually live much longer than mammals of the same size.

Comparing animal longevities is tricky. You can't just compare the ages at which pet birds die with the life span of laboratory rats. Pet birds often die of contagious diseases, while lab rats may be kept in sterile conditions, free of disease.

One solution is to compare the *maximum known longevities* of the entire species. Following this rule, we would compare the age at death of the longest-lived gray parrot with the age at death of the longest-lived rat. Or the age at death of the longest-lived ostrich with the age at death of the longest-lived pig.

The justification for this is that biological maxima are fairly stable. Consider the fastest 100-meter dash at Olympic trials around the world. The fastest times are relatively stable compared to the average 100-meter dash times at company picnics or family get-togethers. You have Uncle Bob wheezing his way to the finish line, and his time could be five times longer than the fastest time. Or he could be in great shape, and his time could be close to the fastest time. Taking the best record throws out all that distracting variation.

The other thing we have to be careful about is *body size*. Larger animals live longer than smaller animals of the same kind. Great apes, cows, elephants, and whales live longer than shrews, mice, rats, and rabbits. This is an interesting fact by itself.

Let's deal with the body size issue, before we return to the bird question. *Why* do larger animal species live longer? One conceivable answer is that merely being larger makes living longer inevitable. Larger animals might have greater reserves of key cells, or they might be more efficient metabolically. Very small animals, like shrews and humming-birds, are famous for the large amount of metabolic work, per ounce, that they have to do just to stay alive.

There are a variety of problems with this answer. First, larger dog breeds die before smaller dog breeds. Great Danes, Saint Bernards, and mastiffs all tend to die before toy poodles, spaniels, and whippets. Second, one of the largest birds, the ostrich, dies at earlier ages than almost all other birds when kept in captivity. Even in humans, there is a smattering of data that suggests that tall men tend to die before smaller men. Bigger isn't better, as a matter of raw physiology.

Why then do elephants outlive dogs, which outlive rats? Larger animals tend to have fewer predators, among other ecological advantages. Think of how much less vulnerable a tree is to mechanical damage, compared to a shrub. Larger animals may be less vulnerable to famine, if only because they can carry a lot of fat around more readily. Think of bears, or the actor John Goodman. These ecological differences allow larger species to live longer in nature.

Now the argument gets interesting. If a species lives longer in nature, the force of natural selection will be increased at later ages. Larger organisms can reproduce at later ages because they are more likely to be alive then, so the force will remain high at later ages. This fosters selection for genes that will tend to keep the larger alive still longer. Thus the longer-lived elephants, oaks, and whales.

But there is nothing inherent about being bigger that is good. The giant dog breeds get no survival benefit from their size. Indeed, the increased burdens that come with their size probably kill them earlier. Physiology by itself is not the key. It's physiology *plus* ecology *plus* evolution that tells the story.

So we have to compare organisms that are alike in size, because the effect of size may otherwise swamp any other effect. When we do that, interesting patterns emerge. For instance, who tends to live longer, aside from size? Flying animals first and foremost, compared with other animals. Small birds live longer than mammals of the same size, with one notable exception: bats. And bats fly. Again, is this a physiological benefit? Does the forelimb workout of flight make the difference, like the arm movement of orchestra conductors who seem to live almost forever?

When flies are given smaller enclosures, they fly less. When they have larger enclosures, they fly more. When their wings are clipped off, they don't fly at all. The less they fly, the longer flies live. Flight is not inherently beneficial.

Evolutionary reasoning gives an answer to the riddle of flight in aging. Flying animals will be better at escaping predators. They will also be able to fly some distance to find food, whenever there is a local famine. For these reasons, flying animals tend to live longer in the wild. Because they live longer, they also have more chance of future reproduction. Therefore, all else being equal, the force of natural selection favors the continued survival of flying animals more than those that can't fly. Thus evolution produces parrots that can easily live 60 years, if not longer, while rodents of the same size that don't fly die in six years.

Larger body size and flight do not exhaust the circumstances that lead to the evolution of long life. Shells and fire-resistant bark might explain why turtles and redwoods, respectively, live so long. Take turtles. Your typical turtle has a maximum life span between 60 and 80 years. Other reptiles, such as lizards, snakes, and crocodiles, some of which are quite large, have maximum life spans around 10 to 30 years. What could be a better defense against predators than a thick shell? Interestingly, soft-shelled turtles are at the lower end of turtle longevities, even under lab conditions.

The same principle operates when we compare the longevities of thick-shelled clams with the longevities of squid and snails, all members of the mollusk group of animals. The thick-shelled species live longer than the species with thin or no shells. Having a thick shell is an evolutionary anti-aging device, because it reduces mortality and thereby increases the force of natural selection at later ages. The force of natural selection dotes on those who live longer in the wild, helping them to live longer when they live in zoos, and thus die from aging.

Natural selection also favors those who are fecund when they are older. When later reproduction is common, natural selection will push the soma to survive long enough to reap the late harvest of offspring. Two groups show this pattern dramatically: aquatic animals and trees.

Many fish continue to grow during adulthood, increasing substantially in fecundity. Some of these fish species live a very long time, and some scientists think that they have virtually negligible senescence. Lobsters can also grow to large body sizes, increasing their fecundity as they go. Lobsters are thought to live longer than any other member of the arthropod group made up of insects, spiders, crustaceans, scorpions, etc.

Trees dominate among the longest-lived terrestrial organisms. Trees are prolific reproducers at advanced ages. They sometimes cover sidewalks with their seeds. Trees add branches and fruiting structures as they grow, increasing their fecundity as they do so. Trees as a group include species that live thousands of years, such as bristlecone pines. Trees that live more than a hundred years are not exceptional. In terms of longevity, trees are the truly long-lived species, and almost all other organisms, plant or animal, are short lived by comparison.

Yet fecundity does not increase life span by itself. Rather, more fecundity usually goes with earlier death. Again, the benefits of increasing fecundity are not physiological, but evolutionary. Even though lower fecundity is better for the survival of most organisms, increased fecundity later in life gives the better evolutionary outcome for the aging of the species as a whole. This happens because increased fecundity late in life gives more of a reproductive future for older organisms. The more reproductive future, the greater the force of natural selection will be. The greater the force, everything else being equal, the more likely it is that evolution will produce an increase in longevity. The physiological effect of increasing fecundity and the evolutionary effect are opposite.

The patterns of longevity among different species are crude and contingent. Some turtles don't live that long, but then they have soft shells.

Some of the other reptile species live a long time, but these species are venomous snakes, and therefore have a defense against predators that is comparable in effectiveness to a shell. Many bat species hibernate, which some have offered as the reason for their greater longevity compared to rats. The qualifications that must be considered when comparing life span in different species go on and on.

One way to escape this problem is to compare organisms with similar evolutionary backgrounds. For example, we might compare mice with rats, since both are rodents. The most extreme form of this comparison uses individuals of distinct types from the same species. In such studies, the evolutionary background is virtually identical. The only comparison like this for most animal species is between males and females. Sometimes even males and females are remarkably similar biologically. The monogamous bird species can be fiendishly hard to sex. In such species, there may not be enough "life-style difference" to make the comparison of males and females interesting, because their ecology, and thus the evolution of their aging, is identical.

The more interesting comparison is when males and females lead very different lives. Maleness appears to be different from femaleness with respect to aging in many species. Most often, males die sooner than females. This is not related to the small "degenerate" Y-chromosome, which makes men male in our species, because females are sometimes the sex with the smaller chromosome in other animals. However, this pattern of early male death is by no means universal, or even profound. In my laboratory, we can switch the life-span sex-difference back and forth by manipulating access to food and new mates. As a general rule, we find that well-fed females die sooner than males, but males who are given a sexual cornucopia burn themselves out on copulation, and die quickly. Abstemiousness in eating and reproducing generally brings the greater life span. (We will talk more about diet later.)

The evolution of male inferiority in longevity may arise from differences in the mating success of males and female animals. Fertile females rarely have much difficulty getting mated in most animal species. Males, on the other hand, are often unsuccessful in mating. Sometimes it is younger males who have less success, because they don't have a territory or they haven't established dominance over other males. This pattern is common in animal species where males compete in a complex social system, as mammals often do. A few older red deer males, for example, may monopolize the sex lives of a group of fertile females, leaving younger males with few opportunities to reproduce. (Think of

Hugh Hefner.) This social system favors the evolution of greater male longevity, because male reproduction occurs more often later in life, compared to female reproduction. Typically, males in species with these ecological patterns live longer than females.

But most animal species don't have such elaborate mating competition. They don't have hierarchies or territories that older males can manipulate to their advantage. There may still be competition between males for access to females, however. When this competition revolves around physical robustness, older males are likely to be much less successful than young adult males. This may leave older males cut out of the reproductive process, compared with older females. Once the older male has little chance of successful reproduction, natural selection may abandon him, the force of natural selection acting on survival dropping to zero earlier in males than females. Genes that have sex-specific effects would then evolve according to whether they benefited males or females. Genes fostering male health later in life wouldn't be favored compared with genes fostering female health late in life. Males should then die sooner than females, even under good conditions. This should be the common difference in longevity in species that do not have male reproductive access improving with age, when there is any difference at all between the sexes.

Even better than the comparison of males and females within a single species is the comparison of different castes within the sexes of social insect species: termites, ants, wasps, and bees. The best-known example is the domesticated honeybee. In honeybees, most of the active members of a hive are female. The males, called drones, buzz around and try to mate with the queen. However, she has too much to do to hang out with the drones. She usually gets all the sperm that she needs during her maiden mating flight, in which she mates with a succession of males, accumulating vast numbers of sperm.

Most of the tasks of the hive are carried out by the female workers. Honeybee workers are sterile. They have evolved to devote their lives to the rest of the hive, and the reproduction of the queen in particular. Meanwhile, once a hive has been built, the queen devotes her life to producing eggs. The life-span difference between queen and worker is enormous. In the growing season, a worker might die a burned out wreck within one to two months. The queen can easily live for five years. However, even this figure is probably misleading, because the other hive members will usually kill the queen once she has run out of sperm.

This is one of the larger proportionate differences of life span within one sex of a single species. It is entirely nongenetic. Queens are produced when workers feed them royal jelly during their larval growth. Royal jelly is a part of the bee's hormone signaling system.

But you can't increase your human life span 30-fold by consuming royal jelly. The royal jelly triggers a developmental pathway specific to the honey bee. *Both* honeybee queens and honeybee workers have the genetic potential to develop into queens when they are eggs. But the workers are hormonally directed to become workers with a short, sterile life, while the queen is shunted onto a track that leads to a much longer life, and considerable fecundity.

It is very important that the queen is vastly more fecund than a worker, yet the queen lives longer too. This occurs because the hive members have been selected to produce queens who will live a long time and workers who can be used up quickly. There is no reproductive trade-off between fecundity and longevity in honeybees. This illustrates the potential that evolution has to achieve long life spans that are also productive.

However, transferring the lesson of remolded aging from honeybee to human is perilous. As mentioned, royal jelly alone is unlikely to do it. It is just the trigger for a series of physiological realignments in the bee. Intervening in human aging using royal jelly would require developing the entire *hormonal response system* of the bee for human use, the response system that makes queens reproductively active, larger, and more durable. So don't put royal jelly on your toast in the morning. What honeybee queens achieve provides us with an early clue concerning the feasibility of postponing human aging. Evolution is showing us the vast potential for altering aging physiologically within one generation, without changing genetics. If we were to accomplish the same feat for human aging, we might live thousands of years. Any such accomplishment would take centuries, but then so did the development of modern printing and publishing. We will return to this somewhat interesting topic in a few chapters.

10

Deadly Serendipity

At Dalhousie University in Halifax I began to study my second set of Methuselah flies, the flies I started breeding in 1980. In the winter of 1982, I collected longevity and fecundity data every day. Unlike my graduate studies at the University of Sussex, I received significant help from a technician, Anne Coyle. Anne had migrated to Nova Scotia from Scotland, but usually managed to hide her Scottish accent unless she lost her temper, when it returned full force. That winter of 1982 was dreary, the sidewalks covered with a treacherous combination of ice and water, making the walk to the university a perilous balancing act. I considered myself fortunate not to be driving a car.

When I got to the laboratory one fateful morning, Anne was at her desk with her face buried in her hands.

I came up to her and asked what was wrong, because she looked like she might have been crying. Anne pointed mutely to one of the incubators, but didn't look up. I went to the incubator and opened the door.

Within was a box with flies inside it, the kind of enclosure that fly researchers call a "fly population cage." Cages provide environments in which thousands of flies can live out their lives: flying short distances, fornicating, and laying eggs. Fly cages are usually supplied with food and reasonable ventilation. Mine are also equipped with clear plastic panels that allow us to see the flies inside without opening the box up.

I noticed immediately that the cage in the incubator didn't have any food, and inferred that Anne had failed to supply the flies with food the night before.

I took the cage out. "It's not a problem," I said. "Almost all the flies seem to have survived, even without food."

I put the cage down on the lab bench.

Anne got up, looking at me quizzically. She didn't say anything. She just pointed at another cage on the bench, one that she must have taken out of the incubator.

I looked at the other cage. It didn't have food either. Then I noticed that most of the flies in that cage had died.

I looked back at the cage that I had taken out of the incubator. It was marked "O", the letter not the numeral, the coding for the Methuselah flies. Most of those flies had survived.

The cage that Anne had taken out was marked "B", the code for normal flies. Those flies had died.

I guessed that the two cages had been denied food for the same length of time. With that assumption, the conclusion was obvious. The longer-lived flies were superior at resisting acute stress. Postponed aging was associated with greater resistance to stress, I realized.

I began to smile. We had a new way to study aging.

Anne looked at me like I was barmy.

It was one of those moments of supreme clarity, almost exhilaration, such as a detective might experience figuring out a murder, or a novelist might have in finally discovering a workable plot. Scientists only get a few such moments in a career, and I am sure we look idiotic when we have them.

What Anne had uncovered was a window into the greater longevity of the Methuselah flies. Stress resistance was correlated with the postponement of aging. If we documented this in properly designed experiments, we could reveal what the postponement of aging was all about, what its physiological *meaning* was. At first blush, if all this could be applied to humans, postponed or slowed aging meant being more robust, not dribbling away years in an enfeebled condition. More eighty-something tennis players, fewer bedridden people in institutions.

Still there was more. It is very hard to find youthful characteristics that predict eventual longevity. Yet such indicators would be invaluable in the study of aging. In Anne's serendipitous experiment, the normal and Methuselah flies that she had starved were young. The ability to resist stress as a young fly is much easier to measure than total longevity. Stress resistance takes a few days to assay. Total longevity takes three months. I hoped to accelerate my experimental program five- to tenfold, using the ability to survive deadly stress as a surrogate for delayed aging.

And yet more was apparent. I now could take a new approach to the physiology of aging. In the past, the physiology of aging was studied by comparing young animals with old animals, or normal animals with those that died young. Anne's experiment was a pilot for the physiological comparison of normal and Methuselah flies, focusing on functional differences that made the Methuselahs live longer. We could look deeper, to find the biological causes of increased stress resistance, and thus the biological basis of staying alive longer.

I didn't have time to take this whole project on by myself. I was an assistant professor in 1982, with a world of newfound responsibilities. Fortunately, I was able to attract a postdoctoral fellow to my laboratory to take on the project, Philip Service. Phil is ten years older than I am, with a professorial look that is both bespectacled and toothy. Whether it's giving a class, performing an experiment, or writing a scientific paper, Phil is organized, smooth, and just too damn perfect to be a human being.

This made him ideal for the task I gave him: figuring out the physiology of life and death in my fruit flies. It would take patience, meticulous experimental design, and an open mind. He had almost nothing to go on besides Anne's inadvertent experiment.

Phil had to make serendipity into science. The first question was which lethal stresses did the Methuselah flies resist better? Why had they survived starvation for more than 12 hours?

This was not as obvious a question as it might seem. The food that we give our flies is also their source of water. The food is the main thing that defines the environment for the lab fly. So Phil had to cast his net broadly to figure out how the stress resistance of the long-lived Methuselah flies was increased.

Phil compared the normal and Methuselah flies for survival under dry conditions. This was his first success in my lab. He found that the Methuselah flies lived longer under dry conditions than the normal flies did. This showed an association between the ability to resist drying out and postponed aging. We didn't know what that meant yet, but it was a clear result.

Phil's next success came with the fly's ability to survive starvation, also an obvious character to try after Anne's accidental experiment that deprived the flies of food. We controlled for resistance to drying in our starvation experiment by supplying the flies with water during starvation. If we hadn't supplied them with water, the flies would have died from drying out long before they starved to death. The Methuselah flies survived total starvation longer than the normal flies. We now had two characteristics for which the Methuselah flies were superior, characteristics that could be measured in young flies as well as old flies: resistance to starvation and resistance to drying out.

Phil's next move was to see if he could figure out why stress resistance was enhanced in the Methuselah flies. One of his best ideas was to examine the effects of age and gender on stress resistance. Resistance to drying out fell with age in both males and females, the Methuselah flies being superior to normal flies at every age. Because a decline with age is the most common pattern of aging for most physiological functions, this was not particularly revealing. Parallelism between such declines is not clearly informative as to physiological causation, because the decline in the force of natural selection can produce such parallels between multiple characteristics without any physiological connection between such characters.

The pattern of starvation resistance differed, which was exciting. Older males survived starvation just as well as young males. Normal flies were inferior to Methuselah flies throughout adulthood, in the male. In females, starvation resistance *rose* with age, a very unusual pattern of aging. Both normal and Methuselah flies showed this pattern, but Methuselah flies were again superior at all ages.

The aging of starvation resistance was distinctive. Only a few physiological characters depend on sex and age in the same way, particularly fat content. Could fat be a key to the evolutionary control of aging?

When Western journalists took trains through the Ukraine after Stalin's mass-starvation program, in town after town they found that almost the entire population was made up of older women. This would make sense if adult women tended to accumulate more fat compared to children and men. Then they would be the only ones who would be likely to survive mass starvation. Would the superiority of our longer-lived flies come from similar causes?

Phil looked at the fat content of the flies. Fat content did indeed increase with age in females, but not in males. He also found that the longer-lived Methuselah flies had more fat at each particular age than the normal flies. This correspondence suggested that fat content was the major determinant of resistance to starvation. Furthermore, it implicated fat in the increased life span of Methuselah flies, since all the Methuselah lines had more of it.

An additional possibility remained. Reduced metabolism would also explain increased starvation resistance. A common finding in stress studies is that reduced metabolic rates are associated with increased stress resistance. This is intuitively natural, because a slower metabolism will use up fewer calories and less water, helping the organism to survive both a lack of food and a lack of water. Many dieters try to manipulate their metabolic rates to increase their loss of fat. Phil checked metabolic rate, and found that it did not match the pattern of starvation resistance with respect to either age or type of fly.

To this point, we had two types of stress resistance that were associated with postponed aging, and Phil had in turn strongly connected fat content with one type of stress resistance, starvation resistance. These results were promising, but a problem was that we had not shown directly that increases in resistance to starvation or desiccation would give increased longevity.

In 1987, I emigrated to the United States, specifically California, from Halifax. I was in need of more good colleagues and, like all experimental scientists, more research funding. I may have been running away from a city in which I had experienced marital disaster, as well.

The reception I received in the United States was a challenge. There was a lot of resistance to the evolutionary theory of aging by American scientists; few appreciated my work from the 1970s. This was brought home to me forcibly when I went to a Gordon Conference on the biology of aging in February 1988, at Ventura, California. Gordon Conferences allow the elite of each scientific field to compare notes, like the elders at gatherings of tribes. I had never been to a mainstream American meeting on aging before Ventura.

My session chair, George Martin, asked that I come to Ventura to speak on common mechanisms of aging among species. I don't think that George knew what he was getting into when he asked me to speak. My talk was scheduled for the afternoon, when most of the East Coast scientists should have been at the Ventura beach, enjoying a February respite from winter. I expected only a few sleepy listeners.

Despite the hour and the distractions of volleyball, I was surprised to see a considerable crowd. In the small amount of time I had, I first outlined the concept of the force of natural selection and then showed how it could be used experimentally to produce longer-lived fruit flies, or indeed longer-lived animals of any kind. Then I said that every nonevolutionary theory of aging was therefore likely to be fundamentally wrong, even though evolution could generate specific physiological problems in older animals, problems that might be mistaken for fundamental causes of aging. Evolution was the ultimate controller of aging, I argued.

This provoked a barrage of attacks. The assembled elders of American aging research insisted that what I had done was not legitimate science. Again and again they came at me. I remember one fellow, bald with a big black moustache, who questioned me repeatedly. He got so red that it seemed as though he might expire from a heart attack at any moment.

After more than 40 minutes of these questions, an eternity by academic standards, George Martin called things to a halt and let the next speaker, Caleb Finch, go on. Finch had watched my predicament from the sidelines, greatly amused. We have been friends ever since.

It is one thing in science to have produced a definitive result, even one that your immediate peers recognize. It is quite another for the meaning of that result to sink in with the full range of scientists in your field, especially if that field is broad, as the study of aging certainly is. The scientific revolution that the evolutionary approach offers has not yet been fully absorbed by scientists who study aging. But the universality, the preemptive status, that physiologists and molecular biologists used to claim for their theories of aging is now rarely asserted. They just don't want to tango with me at scientific meetings anymore. Still, it was fun going up against them while it lasted.

When I moved to the University of California–Irvine, I got the resources I needed to test the impact on longevity of increasing stress resistance. The most important of these resources was the student body of the School of Biological Sciences at UC–Irvine. As my lab filled up with students interested in research projects, I got them to select for increased stress resistance in the flies, to see if it would increase longevity.

We did this experiment the old-fashioned way. We killed a lot of flies. To select for increased starvation resistance, we kept thousands of flies in cages that had a source of water, but no food, until 80 to 90 percent died. The survivors were then given food, allowed to recover, and left to reproduce normally. We selected for starvation resistance over dozens of generations.

Selection for resistance to drying out was similar. Flies undergoing selection were dried using chemical desiccant. They had no water and no food. The flies were dried out until 80 to 90 percent had died, and then the survivors were allowed to recover and reproduce. We started to select on these flies in 1988, and they endured selection for more than 15 years.

Both selection experiments increased stress resistance from generation to generation. This response to selection was genetic. Flies from selected lines showed an improvement in stress resistance even when they were taken off selection and raised in a normal environment. At the same time, longevity increased from generation to generation. This showed that stress resistance is one of the things in fly evolution that controls aging.

What I needed next was a way to figure out how this worked. What was the physiology that explained starvation resistance in our flies? If we could exclude variation in metabolism, could starvation resistance be explained just by stored fat calories? I recruited two physiologists to answer this question: Tim Bradley, a professor in my department at University of California–Irvine, and Minou Djawdan, then his postdoctoral fellow. They looked at the full range of stocks that we had created by selection: flies that lived longer, flies that resisted starvation, flies that resisted drying out, and normal flies. They looked at both fat and carbohydrates. While stored fat was a fairly good predictor of the ability of flies to survive starvation, the best predictor was the total of all stored calories, whether fat or carbohydrate.

This gave us a simple conclusion to our story. Flies live longer when they have more stored calories, even if they have access to food. When flies reproduce more, they deplete their stored calories. And then they die.

What about the physiology of resistance to drying out? Phil Service found that desiccation resistance was increased in the longer-lived Methuselah flies, but he hadn't found any clues to what determined desiccation resistance. Fortunately, desiccation resistance is a classic problem in insect physiology. A whole team of insect physiologists set to work on the problem of aging and desiccation with my flies: Tim Bradley, another faculty colleague Allen Gibbs, Minou Djawdan again, Adrienne Williamson, and Donna Folk. Even Adam Chippindale got involved, we were that excited. They looked at a grab bag of possible mechanisms that might enable flies to increase desiccation resistance: metabolism, ability to endure water loss, stored calories, activity, pattern of ventilation, and so on. Years of fastidious work came down to one simple thing: how much water a fly has stored, and how fast it loses that water, determine resistance to death by drying out. Once their water content falls below a threshold, flies fall over dead. This is completely unsurprising, in hindsight, but you have to demonstrate the unsurprising in science using experimental data. You can't take it for granted.

But why should the loss of water be associated with fly longevity? Water is required for survival. Many organisms have adaptations that conserve water, especially organisms that don't live in water. Older flies tend to lose pieces of their bodies as they get older. Flies that have lost the tips of their legs are not only mechanically handicapped, they also have suffered a breach to their cuticle, the exoskeleton of the fly. One of the functions of the cuticle is to keep fluid inside the fly. As the cuticle is mechanically ruptured with age, hemolymph—bug blood—will be lost, and with it water. Thus it becomes harder at older ages for the fly to retain water. Impaired walking and flying will reduce water intake. The older fly faces an increasingly difficult problem with its water metabolism. Having more stored water may help the older fly survive impaired water metabolism. There are still some gaps in this story, but as of this writing we are doing experiments on the role of cuticle damage in the control of aging and desiccation resistance.

Fifteen years of targeted selection and physiological work taught us something simple: retaining calories and water helps a little tropical insect live longer. We had discovered some of the physiological means by which evolution controls aging in insects.

This work provides two useful lessons for human longevity. First, you can figure out the physiology of slower aging. Second, this physiology can be simple. The prospects for improving the quantity of stored calories or stored water are a lot better than the prospects for manipulating some complex piece of molecular biology, like error catastrophes.

A skeptic might complain that a small insect's aging might be controlled by such simple things as calories or water, but surely the aging of a large, complicated organism like a human being must be based on more subtle biochemical processes. This is a seductive prejudice.

But seduction is not truth. Some decades of research have shown that something as simple as caloric intake might be an important control for human aging. That is the theme of the next chapter.

One Can't Be Too Rich or Too Thin

Our work on starvation in fruit flies left me with an unfulfilled intellectual hunger concerning the relationship between fat, calories, and aging. I was particularly curious about the relationship between metabolism and survival. Could the entire aging process be explained in terms of a few simple physiological principles, such as the consumption and use of calories? This made sense in terms of basic biological concepts, but I have always distrusted the intuition of biologists, my own included. Animals do not seem to read our scientific publications. Instead they are more than willing to evolve in ways that we can't predict.

Even though I aspired to find a scientific integration of nutrition, evolution, and aging, I doubted that any such integration would ever be found. I would be wrong about that. Bear with me as I describe the circuitous route by which we have figured out the significance of diet for aging. I personally guarantee that you will find our conclusions interesting and, for some, of practical importance in your daily lives.

The relationship between aging and diet has been studied for some time. The first experiments date back to the 1930s. The primary data come from caged rodents. Rodents in experimental cages have nothing to do but eat, excrete, and sleep, unless they have exercise wheels. Most don't have wheels. They are usually housed singly, so they can't even have sex. Instead they eat. And eat, becoming quite fat, usually obese. This happens because they are given an unlimited supply of food. But if caged rats or mice receive 20 to 40 percent fewer calories than they eat when their appetite is given full sway, they live longer by about the same amount, 20 to 40 percent. (The quantitative effect varies between experiments.)

In the early days of these experiments, they deprived rodents of food when they were still immature, leaving the rodents with delayed maturity. This led me to think that reduced calories might increase total life span merely by stretching development, the period before adulthood. This effect is well known from research on insects and other animals. Reducing food calories stretches out the length of time for the development of fruit flies, for example. This is just brute fact, whatever the explanation. If the same effect occurred in mammals, then part of the increase in life span in rodents that are fed less food would have nothing to do with aging.

When I moved to the United States in 1987, I learned that modern-day research on diet and aging had improved considerably since the 1940s. In the 1980s, rodent experiments made it clear that the effect of reduced nutrition wasn't just stretched development. Calories could be reduced *after* the sexual maturation of rodents, and it would still increase life span. Reduced caloric intake can be started fairly late, even in middle-aged rodents, and longevity will still increase, though not by as much as when reduced food intake starts earlier.

Of great interest to me, though of little interest to medically oriented biologists, was the fact that reducing dietary calories seemed to work in many other species, including invertebrate animals. This suggested that reducing dietary intake should increase life span in insects as well. Unfortunately, the published experiments on diet and aging in fruit flies, the experimental system I knew best, were a mess. Sometimes the effect was there. Sometimes it wasn't. My guess was that bad technique might have been to blame for this inconsistency.

The effect of caloric reduction on life span was the most significant discovery of aging research before the evolutionary manipulation of aging. It was the only way that had been found to make mammals live longer, excepting perhaps castration, an intervention with little potential as a medical solution to the problem of aging. On the other hand, the increase in longevity with reduced calories wasn't an achievement of scientists. It was something that their experimental animals did. There was no scientific theory that led us, step by logical step, to expect these dietary interventions to work. Instead, it was a common finding that, if you gave animals less food, they sometimes lived longer. This is even one of the nuggets of folk wisdom about human aging, dating back centuries. People who ate abstemiously were widely supposed to live longer than those who did not. But more about that later.

When I first started work on the evolutionary connection between diet and aging, it was more a puzzle waiting to be solved than a finely honed piece of scientific reasoning. I hoped to change that. I wanted to make sense of the impact of diet on aging. My tool for doing that had to be the fruit fly. Rodent dietary experiments took several years, too long for me. I wanted to move quickly.

My first objective was to reappraise the previous dietary experiments with fruit flies, some of which had increased longevity, some of which had decreased longevity. I thought that a positive effect on life span would require an intermediate level of nutrition, because too little would leave the flies starving to an early death.

I set a team of undergraduates to work on the problem. They gave some flies unlimited amounts of food, and other flies fractionally reduced amounts of food. The easiest way to do this was to vary the amount of yeast paste that was spread on the surface of the Jell-O–like food that flies are given in my lab. That yeast paste was the food of choice for the adult flies. Out in nature, our fruit flies actually don't eat fresh fruit. They eat the slime of rotten fruit, slime that has lots of fungus and bacteria. Our yeast paste was like that slime, because yeast are fungi. No accounting for taste.

Lots of yeast paste gave the typical adult longevity, about 30 to 40 days in normal flies. Significantly less yeast paste, say one-tenth the maximum amount, gave a slight increase in adult longevity. Giving the flies no yeast paste at all actually reduced longevity compared to the effect of low levels of yeast paste. So dietary reduction didn't always increase fly life span. You had to get the level of nutrition right. That might explain the great disparities in the effects of diet on fly longevity in the published experimental literature.

This was the starting point. We had dietary restriction working in our flies, controlling it using yeast paste. We had established the fact that dietary restriction could give increased longevity in our flies, under the right conditions. That meant that we could now start figuring out *why* dietary restriction sometimes produced increased longevity.

The next thing to do was compare my normal flies with their longer-lived cousins, the Methuselah flies. I was intrigued by the improved starvation resistance of the Methuselah flies, described in the preceding chapter. One possibility was that the Methuselah flies had altered feeding. Could I have selected simply for flies that ate less, like a group of ascetics living in a desert? If so, then imposing reduced diets on the Methuselah flies and the normal flies should increase the longevity of the normal flies to the level of Methuselah fly longevity when the Methuselah flies had abundant food. By contrast, reduced food should, on the basis of this theory, leave Methuselah fly longevity unchanged or perhaps even reduced to a level *below* that of normal flies, if their conjectural tendency to eat less got them into trouble when less food was available.

At this point, I handed the project over to Adam Chippindale and Armand Leroi, my new Canadian graduate students. It was the 1989–90 academic year at the University of California–Irvine, and we were setting off on a sequence of experiments that would lead us into strange territory. Adam and Armand, who led a lab group known as the A-team, were the advance guard.

The first experiment they did compared the long-lived Methuselah and normal flies with and without reduced diets, using the nutritional levels that our preliminary experiments showed would increase longevity. In a much larger experiment, the A-team again found a beneficial effect of reduced food in the flies: about a 10 to 15 percent increase in adult longevity. This effect happened in both normal and Methuselah flies, in males and in females. This showed us that dietary manipulation worked well in both normal and Methuselah flies. This was evidence against the idea that the Methuselah flies lived longer because they were restricting their food intake. Forcing the Methuselahs to eat less made them live even longer.

To this point we had shown that dietary restriction worked in flies and we had shown that the longer-lived Methuselah flies were not just dieters. But what sense could we make of the intersection between longevity, diet, and resistance to starvation?

In the same experiment in which we imposed dietary restriction, we also monitored egg laying. My intuition was that reproduction had to be key to the impact of diet. Reduced caloric intake was known to impair fertility in rodents. In some rodent dietary experiments, ovulation ceased in the group with reduced access to calories, as it does in some women with anorexia nervosa, the body-image disorder in which victims greatly restrict their food consumption.

This was a small detail to a medical researcher, but a clap of thunder for an evolutionary biologist. Reproduction is indispensable to Darwinian fitness. If dietary change reduces reproduction, the effect on reproduction has to play a central role in figuring out the evolutionary interaction between diet and aging. No theory for the relationship of diet and aging can neglect reproduction.

The A-team found that the lifetime output of eggs was greatly reduced on a diet with less yeast paste. In fact, it was reduced far more than longevity was increased by restricting food. This made the impact of diet on aging easy to understand. There is abundant evidence showing that reproduction tends to kill the adult, more reproduction resulting in less longevity. If dietary reduction reduces reproduction, increased life span is to be expected to result from a restricted diet. This is just another case in which reproduction trades off with adult survival.

Now the trail was getting hot. We had already found that starvation resistance was increased in the longer-lived Methuselah flies. The Methuselah flies also had reduced early fecundity when young. And starvation resistance appeared to follow fat content, with fatter flies resisting starvation better—no great surprise. If all this were true, perhaps diet controls a shunt that diverts calories either to fat stores or to eggs? If so, then we should see an immediate change in both fecundity and starvation resistance when food intake is varied.

The A-team set up a series of yeast-paste dilutions for adult flies, from almost no yeast in the paste to a rich, yeast-laden paste. They fed these different pastes to groups of flies over a series of days. What they found amazed me. If they took females that had been receiving abundant yeast and switched them to low yeast, within a few days their egg laying fell to low levels, while their starvation resistance rose. Paradoxically, as the fly food intake fell, they were apparently putting away more calories in the form of stored fat. They could do this because they were shutting down reproduction.

This may be depressing news for women who use dieting to lose weight—if insects are a reliable guide to human physiology. There is some evidence that these findings do indeed apply to humans, in the increasing difficulty that women have losing weight after each round of dieting. While we do not know all the relevant human physiology, there is some evidence for a short-term impact of the human diet on our reproduction. In particular, substantially reduced food intake impairs fertility in people who are not overweight to begin with.

In our fly experiments, the converse was true as well. Within a few days of improved nutrition, female flies increased their output of eggs and decreased their starvation resistance. Furthermore, there was a simple trade-off between starvation resistance and egg laying, at intermediate food levels. More food led to more eggs and less starvation resistance, the degree of starvation resistance following the quantity of fat in the female body.

This trade-off arises from the female fly's abdomen. In fruit flies, the abdomen is a large fraction of the body of the female, about one-third, depending on her reproductive activity. Within the abdomen there are two main structures: the ovaries and the fat body. The ovaries of course make eggs, just like human ovaries. But in fruit flies the eggs are vastly larger than human eggs, and far more are released per day. Because of this, fly ovaries are enormous relative to the size of the whole fly. The fat body stores fats, as its name implies. It also does a variety of other things, from processing nutrients for the eggs to producing the limited immune response of the fly. In some ways, it is the "liver" of the fruit fly. But for our purposes the important thing is that the fat body and the ovaries are organs that use a large fraction of the calories that the female fly consumes.

An interpretation was easy to find. Dietary restriction worked in our flies because of a shift in resources from ovaries to the fat body when nutrition was reduced. With more calories stored in the fat body, the fly was better able to survive. The Methuselah flies lived longer in part because they had more calories stored as fat.

Why would older flies need more stored calories to survive? Older flies are falling apart, literally, as I described in my discussion of the stress of drying out. Pieces of their wings and their legs fall off. They find it harder to move about as they get older, even in the protected environment of the laboratory. Therefore, as they age they probably can't feed as efficiently.

Think of an old man in the supermarket, taking forever to choose his food, pay for it, and get it to his car. Later, in his apartment, he has to struggle with his stove to cook his meal. His dentures may make chewing difficult. This is not an argument in favor of human obesity. It is the simple point that the loss of mobility later in life will make feeding harder. Indeed older men, especially those without a woman to take care of them, tend to be poorly nourished and tend to lose weight.

There is an alternative to this analysis, which is based on metabolic rate. It is a longstanding criticism of dietary restriction research that the longer-lived, calorie-restricted rodent has a reduced metabolic rate. We know that reduced metabolic rate in invertebrates produces an increased life span. We've known that for more than 80 years, thanks to some of the earliest and best experiments on aging using fruit flies.

Since fruit flies are small, cold-blooded animals, increasing or decreasing the temperature of their environment correspondingly increases or decreases their metabolic rate. Flies with higher metabolic rates die sooner than flies with lower metabolic rates. Therefore, interpreting the physiology of aging requires knowledge of metabolic rates. What my colleague Tim Bradley and his collaborators have shown is that metabolic rate per unit weight is not a factor in our fly experiments, if you focus on lean weight. Fat can't be counted along with metabolically active tissue, because it just sits there. (Roughly speaking, the same is true in humans. Most of our fat reserves have very low metabolism.)

Likewise, it turns out that reduced dietary intake doesn't reduce metabolic rate per lean pound of rodent. Less food does tend to reduce the total body weight of rodents, because there is less fat, and the metabolic rate of the entire animal is also reduced. But this is just scaling. Shorter humans tend to weigh less than taller humans, but this doesn't mean that they are thinner. Flies that endure moderately reduced food, complete starvation, or desiccation also do not reduce their metabolic rate, adjusted for weight. Total metabolism is not an explanation for the connection between diet and aging in flies and rodents.

What *is* the relevance of this research to human aging? Dietary manipulation makes a variety of animals live longer, especially rodents, but it does not always work. Would it work in humans to postpone our aging substantially?

It is not an appropriate criticism of the importance of diet for aging to bring up the early deaths of concentration camp victims and anorexia nervosa patients. These people are starved to a high degree—they are malnourished. They die sooner because of acute starvation. Flies and rodents in dietary restriction experiments are given access to adequate basic nutrition. They are only denied the additional calories required to reproduce more or to become obese.

The ideal human experiment would use subjects who are nourished scrupulously well, but denied calories. This sounds like dieting, as practiced by millions, but it is rare for thousands of people to follow the identical diet together, much less millions. Research suggests that it is difficult to get us to follow a stringent diet, when we are left to ourselves. We just love to eat food on the side. Before the 1990s, it was hard to come up with evidence supporting the hope that reduced caloric intake would increase human life span.

But providence intervened. In 1991, Roy Walford—then in his sixties—and a group of much younger pioneers were locked into Biosphere II, a sealed research facility in Arizona. Roy was a gerontology professor at UCLA, but a lifelong nonconformist. A wiry man with a shaved head and a prominent moustache, you would think that he had been a musician in Frank Zappa's band. Indeed, I have seen an avant-garde music video featuring Roy that I think Zappa would have liked, had he lived to see it. But Roy wasn't just someone who pushed the envelope. He had a delightfully sly, sometimes extremely obscene, sense of humor too. I can't describe here what you would have seen if you had used the bathroom in his house.

Biosphere II was a trial run for self-contained colonies on the Moon or Mars. It was supposed to be a self-sustaining ecological unit, growing its own food, mostly vegetarian. But the plants in Biosphere II didn't grow as well as originally planned. There was a problem with the balance between oxygen and carbon dioxide. The Biosphereans were stuck in isolation, and there weren't going to be enough calories for them to eat. Roy was able to design a diet that gave adequate nutrients, but it was short on calories. This went on for the duration of their stay in Biosphere II, a period of two years.

The effects of two years of reduced calories were dramatic. The Biosphereans all lost weight, at least 10 percent of their total body weight. Their cholesterol levels fell. Their blood glucose and fat levels fell. Blood pressure was reduced an average of 20 percent. Across a panel of indicators, the Biosphereans had improved cardiovascular health. Roy estimated that their risk of heart attack and other age-related disorders had fallen considerably.

Yet all was not well. Despite being young, and in great shape to begin with, Roy's subjects had little energy. It was hard for them to do the work required to maintain Biosphere II, even to harvest the plants needed for their diet. Some of them became depressed.

It is no surprise then that all the Biosphereans, other than Roy, went off their restricted diets as soon as they emerged from Biosphere II. As they ate more, their blood returned to normal risk levels, with higher serum cholesterol, triglycerides, and glucose. They were also able to resume normal activity levels.

Roy Walford died last year (2004), just shy of his eighty-fourth birthday. About a year before that, I went to a party thrown in his honor by a group of local scientists interested in aging. Roy had severely impaired movement, due to amyotrophic lateral sclerosis (ALS), also known as Lou Gehrig's disease. His apartment was festooned with long ropes that he used to move around. He remained feisty, and took a particular interest in my date, Botum Seng, an exotic-looking Cambodian diva. Roy was notorious for his success with the ladies.

Roy's daughter Lisa and others interested in living longer continue to eat a restricted-calorie diet. They stay in touch with each other by the Internet. They have an on-line manual that shows restrictors how to change their diet so that calories are reduced, but nutritional quality is preserved. Their efforts are unlikely to give data to equal the quality of Roy's short-term, small-group experiment in Biosphere II. But they may nonetheless eventually make the long-term value of human dietary restriction clear.

The National Institute on Aging has begun human trials with low-calorie diets in Louisiana, Massachusetts, and Missouri, but it will be years before these trials will give definitive results. I think that they will need especially good luck to keep the test subjects who live near New Orleans, a city that gives heartbreak to those who try to stay on diets.

One of the ambiguities in the old rodent dietary experiments was that the experiments were conducted under unsanitary conditions in which the rodents often died of infection. In experiments where the rodents were lucky enough to be spared infections, the restricted rodents lived longer. But if a significant infection appeared in the experimental colonies, the quality of the data fell apart. Because of this problem, later experimenters have been careful to maintain their rodents behind barriers that prevent infection by outside pathogens.

The Biosphere II experiment was similar to a modern rodent experiment, because the Biosphereans were isolated. That may have been a factor in the clarity of the data that Roy obtained from Biosphere II— it wasn't complicated by the effects of colds, influenza, etc.

Comparable human data on a larger scale than Biosphere II might come from an island population that has had restricted contact with the outside world. Calories would have to be restricted, but the inhabitants should have had long-term access to varied foodstuffs that provided vitamins and other essential nutrients. At the same time, good vital records must be kept: birth certificates, marriage licenses, and death certificates. Almost no large human populations have met all these conditions. At first, I wasn't aware that any had.

My ignorance on this point was remedied by a trip to Japan in the fall of 2001. I was invited to speak at the Okinawa International Conference on Longevity by David Itokazu, a very solicitous host, as well as an advocate of all things Okinawan. A number of aging researchers had been invited, as well as Andrew Weil, the noted author of books on alternative medicine. As many appreciate, Weil looks like a cross between Santa Claus and Buddha, though slimmer. I found him a source of abundant information on Asian culture and surprisingly skeptical where the American anti-aging movement was concerned. At the Okinawa conference, we met some of the archipelago's centenarians, whose vigor, humor, and character were obvious. Okinawa has one of the greatest concentrations of centenarians on Earth, if not *the* greatest.

The Okinawa longevity conference took place in the Bankoku Shinryokan Conference Hall, toward the north end of Okinawa's main island. The previous year the conference hall had been used by the G-8 Conference, which had left behind framed photographs of world leaders in casual clothes. Most of these photographs were not inspiring where human aging was concerned. But the setting was lovely. The meeting itself was run by the earnest Canadian twins Bradley and Craig Willcox. They have based their careers on the longevity of Okinawans, each in his own way. It is their good fortune that Okinawans are long-lived by any reasonable measure, some older Okinawans sustaining excellent vitality until very late ages. The pertinent issue is why do Okinawans have postponed or slowed aging?

We were given traditional Okinawan meals at the meeting, meals that were an education in exotic foodstuffs. The Willcox twins have eaten the diet for years, and they remain vigorous. But I wasn't so sure. Most of the things I ate were unfamiliar to me as a biologist, and the tastes ranged from the bitter to the nauseating. When we were taken to the urbanized south end of the island, the Okinawan food in the restaurants wasn't so extreme. But it was still difficult to find Okinawan food that had many calories.

During the scientific sessions themselves, I listened to talks about the antioxidant content of the Okinawan diet, its richness in trace nutrients, and so on. Other speakers extolled the Okinawan way of life as a study in serenity. But it was hard to pay attention because I was so damn hungry. Then it struck me. The traditional Okinawan diet verges on starvation. Because of a lack of agricultural productivity, Okinawa did not have much high-calorie food for its inhabitants until the 1950s. Yet the waters off Okinawa afford an abundant supply of highly nutritious, low-calorie foods: various sea weeds, small quantities of crustaceans, and mollusks. The charming Okinawan centenarians that we met were all tiny. Okinawa in the twentieth century might have been the world's largest dietary experiment.

I conferred with my fellow speakers at the meeting. Even those who had little interest in diet and aging agreed with me that the "miracle of Okinawan longevity" might be a low-calorie diet combined with abundant nutrients from low-calorie seafoods.

What does the historical record reveal about the Okinawan way of life? Before the American occupation, Okinawa never had much agriculture, making it chronically vulnerable to interruptions in food supplies from exporters like China and Japan. But the islanders of Okinawa underwent even more severe conditions from the 1930s to the 1940s. World War II, and the military actions that led up to it, interrupted normal commerce between Okinawa and the rest of the world. As American naval superiority tightened a noose around Japan in the later stages of the war, Okinawa was almost completely cut off. Okinawans must have undergone severe caloric restriction, yet they had qualitatively good nutrition and rough quarantine from disease.

If humans respond to caloric restriction with increased life span, Okinawans should live longer. The meticulous demographic data of Okinawa show that Okinawans certainly do live longer. The documentation is now extensive. No matter how the data are analyzed, with respect to average longevity or with respect to maximum longevity, the Okinawans are exceptionally long-lived. Some claim that Okinawans live longer than any other human group. But it is not necessary to agree with this view to come to the conclusion that Okinawans are extraordinarily long-lived.

An interesting aspect of the enhanced life span of Okinawans is that they were forced to build a vast underground network of tunnels for the Japanese military. Then they suffered a spectacularly violent military conquest by the American military, followed by foreign military occupation during which they were second-class citizens in their own country. While this ordeal was not as severe as that of the Jews in Nazi Europe, it was still protracted stress. At the end of the American conquest of the main island of the Okinawa prefecture, there were mass suicides by Okinawans. Yet the Okinawans that survived live much longer than cossetted Americans. So much for the commonplace advice to live longer simply by avoiding stress.

The Okinawa and Biosphere II data, as poor as they are compared to experiments with laboratory animals, provide evidence that caloric restriction can increase the human life span. Therefore, a person who was willing to lead a life of diminished food intake and reduced energy might predictably live longer. However, it is notable that Okinawans don't live to be 140 years old. They enjoy just a few years of increased life, with male and female average life spans of 77.5 years and 85.1 years, respectively. This is a year or two better than the average life spans for Japan as a whole, 76.7 years and 83.2 years, males and females respectively. To achieve this difference in rate of aging, if my interpretation is correct, the longer-lived Okinawans suffered substantial caloric deprivation, some for decades of their lives.

The prospects for a revolution in human longevity from stringent diets are dubious. First, there is little evidence that reducing human dietary calories gives an extension in life span as great as that which lab rodents achieve with reduced diets. Second, most cases of successful human dietary restriction occur when the human subjects have little choice. The youth of Okinawa no longer eat a traditional Okinawan diet. They eat American fast food and other sources of abundant calories, a pattern repeated in East Asian communities around the world, with substantial health effects. For example, Okinawans living in Brazil die 17 years earlier than their relatives who remain in Okinawa, on average. However, this difference may reflect other risk factors, such as smoking and occupational hazards. One effect of the recent shift in Okinawan caloric consumption is that young Okinawans who have grown up with Western diets are much larger than their elders. They also show a more Western pattern of disease, especially earlier cardiovascular disease and cancer. Okinawans traditionally ate a forbidding diet: many servings of fruit and vegetables, with some soy. They saw fish several times a week, and meat or dairy foods hardly at all. Having eaten this diet myself, I find it easy to understand the desertion of the youth of Okinawa to higher-calorie Asian food as well as megacalorie Western food.

But there is a contrary movement of Westerners adopting an Okinawan diet, led by Brad and Craig Willcox. Not only do the Willcox brothers eat the traditional Okinawan diet, and somehow survive, they also write books proselytizing on behalf of the Okinawan way of life. I think that their emphasis on Okinawan tranquility as a source of longevity is spurious, given the mayhem of recent Okinawan history, its invasion during World War II especially. But my guess is that they are dead right about the diet. However, the only chance I see of widespread North American adoption of an Okinawan diet is rigid agricultural policies, like those of Europe or Japan, which drive up the price of food. Then the sheer cost of high-calorie meals would make Americans more abstemious.

George Roth has a different solution to this problem. I got to know George in 1988 at a lobster banquet in Bar Harbor, Maine, held as part of an aging meeting. George has worked on caloric restriction in lab mammals for years. But he is no Roy Walford. Not only did he eat all of his lobster, he ate most of mine too. George is not going to give up his food. George's plan is to figure out the internal controls of the dietary restriction system, and supply medications that would convey the beneficial effect of dietary restriction without giving up on edible food. If George succeeds, he will be one of the greatest benefactors of mankind, as well as a happy gourmet.

Recently George and his colleagues at the National Institute on Aging have focused on the body's burning of glucose, the blood sugar. Their reasoning is that jamming this system might enable people to eat calories that can't be taken up by human metabolism: we could eat lots of food without getting fat. They have had some success feeding a chemical called 2DG to rats. The medicated rats ate unrestricted amounts of food but had many of the same health benefits as the people living in Biosphere II. Unfortunately, 2DG can be highly toxic if taken over long periods or at high doses, making it unsuited to medical use. Nevertheless, George Roth has proven his point. There is the possibility of having caloric restriction and your lobster too.

Reducing dietary calories is a promising entrée into the physiology of postponing aging. It offers some hope for slightly postponing human aging. But it is no panacea. Manipulating nutrition will be only a part of a more complicated approach to the treatment of human aging.

Just how complicated that will have to be is the subject of the next chapter.

12

Many-Headed Monster

It all started with another phone call from Leo Luckinbill. I was still in Halifax, Nova Scotia, and it was about 1984 as I listened to Leo's intense voice on the handset. He claimed to have data indicating that aging in fruit flies was controlled by one, at most two, major genes. If anyone else had said that to me over the phone, I would have dismissed their claim straight away. But Leo was an outstanding experimentalist. His claims I took seriously.

Yet I was shocked. Leo's result was contrary to predictions made by leading evolutionary biologists. Aging was supposed to be affected by many genes, not one or two.

There are two reasons for this. The first is that evolution doesn't overdo it. It won't produce body parts that last too long.

Think of the automotive industry. Henry Ford sent one of his engineers to a scrapyard to find out which parts of defunct Model T's still had some usable life left. When the engineer reported back with a list of the durable parts, Ford instructed his engineers and suppliers not to make those parts to such high specifications. Ford didn't want to waste money making parts so good that they outlasted the rest of the car. This strategy culminated in "planned obsolescence," an ugly tradition in American manufacturing.

Evolution is regrettably similar. If a mutant individual has a liver that will last for 2,000 years, the mutant gene will not be strongly favored by natural selection because the body will die of other causes long before the 2,000 years are up. This would be like the 50-year carburetor. The car is not improved by it, because the rest of the car will not usually last 50 years. Such late-life benefits in animals are either neutral or selected against, if they are genetically associated with deficits in early function. The gene that gives rise to the 2,000-year liver might give rise to more work for the kidneys, so that they fail at 30 years of age. Likewise, a 50-year carburetor might cost an extra $500, money better spent on another car part. Ford realized the value of the simultaneous failure of car parts. Evolution, in its own way, has the same understanding where the aging of body parts is concerned. Henry Ford and natural selection, two indomitable forces of nature.

There is another face to this synchronization of aging. If a single physiological process causes death long before any other, then it is probably killing the animal when natural selection is still acting with some force. Natural selection will favor the postponement of this "leading" process of aging, selecting for genes that forestall that particular kind of deterioration. As this process of delaying death continues, the physiological mechanisms that cause aging early in life should be brought into synchrony with late-acting mechanisms of deterioration.

This happens in the automotive industry when relatively new cars are recalled because of defective construction or design. These cars are repaired free of charge by the manufacturers because their business depends on producing cars that literally hold together for the first few years after they are built. Any defects that arise early in the life of a car will be the focus of strenuous remedies, including recalls.

For these two reasons, evolutionary biologists expect many synchronized physiological causes of aging. Therefore, they have long predicted that all attempts to find the "mechanism," a single all-powerful physiological cause, of aging would fail. When it comes to placental mammals like ourselves, evolutionary biologists have never had to eat their words.

There are some organisms that seem different, however. Some adult insects lack mouthparts. As soon as they mature, they start to starve. This ends their lives quickly. Some species of may-fly are examples of such a curtailed life. Many other organisms have a strong association between reproduction and death, dying within hours or days of their single period of reproduction. Pacific salmon, soybean, marsupial "mice," and bamboo are some examples. Surely these organisms have single mechanisms of death, one might think? Supporting this claim is the huge beneficial effect of castration on life expectancy in species that reproduce only once. Calamitous reproduction is the immediate cause of death for these animals and plants.

But it is notable that castrated soybean and salmon die nonetheless. They don't go on to live for centuries after castration. They live somewhat longer, perhaps years longer in the case of salmon, but the aging process still occurs, leading to death even under the protected conditions of salmon aquaculture.

These findings show that there can be very important, indeed leading, aging processes that normally cause death. But the abrogation of such processes still won't stop aging, only delay it. If the leading mechanisms of aging are stopped from killing an animal, other aging processes will come into play, killing the animal that has enjoyed a partial escape from the aging process.

If there are many partly synchronized processes of aging, then there should be many genes that modulate aging processes in most animals and plants. In practical terms, if one disorder of aging doesn't get you, something else will. In modern-day Western populations, most of our deaths are related to aging. The aging processes are fairly synchronized. Our mortality rates accelerate through most of adulthood, making the mortality rate among individuals in their seventies and eighties crushing.

Try reading the obituaries. Most people who appear there are in their seventies and eighties, with some in their sixties and nineties, leaving aside violent deaths due to combat, accidents, murder, and suicide. In parallel with this acceleration in total death rates, rates of death due to strokes, kidney failure, cancer, and many other disorders accelerate. It is as if there is a many-headed monster responsible for aging, with each pair of slavering jaws trying to keep up with the others as they feed on our bodies.

Yet here was Leo Luckinbill telling me on the phone that only one or two pieces of physiological machinery, one or two genes, were responsible for aging in fruit flies. I had to replicate Leo's finding with my own flies, or refute it. The gene-number project would require a research effort from someone with enormous stamina, because this kind of genetic work needs lots of replication.

The person who took the assignment was Ted Hutchinson, a Canadian from Vancouver. When he came to my Dalhousie University lab in 1983 he was a charismatic hippie fresh from various European romances. He even had a charming accent, the compound of years spent in Quebec. I hated seeing it happen, but I ruined the next five years of his life.

Ted's experiments on the number of aging genes used hybrids. Like hybrids of animal breeds and flower varieties, these hybrids were made by mating very different types of fruit fly. There is nothing unusual about making hybrids in practical breeding or scientific genetics. Sometimes dog breeders cross their pure breeds to create a new breed. But the offspring of hybrids vary a lot more than pure breeds. This happens because hybrids contain genes from both parental breeds, and those genes are inherited at random when hybrids mate with other hybrids. This random pattern of inheritance is called *segregation.*

The pattern of segregation indicates the number of genes that separate two breeds. The key feature is the "lumpiness" of segregation in the offspring of hybrids. Think of porridge. When you cook it in a pot and ladle it out into bowls, it can be lumpy or smooth. Big lumps of porridge indicate that the porridge hasn't been blended during cooking. Smooth porridge is nicely blended. When there are a few important genes controlling a biological process, its inheritance is "lumpy." The offspring receive one or another among just a few important genes in all-or-none lumps. When there are many genes, at the opposite extreme, their effects average out, producing smooth variation, not lumpy variation, in the offspring of hybrids.

In a cross between hybrids of two dog breeds, inheritance is lumpy when it is possible to pick out a distinct combination of attributes from the parental breeds in each puppy. One parental breed might have long ears, the other breed short ears. Some breeds might have black coats, others white. With lumpy inheritance, puppies will have long ears or short ears, black coats or white coats.

When inheritance is smooth, the litters of hybrids would have puppies with ears of about the same length, and coats that are shades of gray, though these characters would still vary. It would be hard to say that one puppy's character came from one ancestral breed, and another character from the other breed. If there are many genes separating the breeds, the puppies of mongrel parents would be intermediate between the breeds.

It took a long time to find out whether the inheritance of fruit fly aging was lumpy. Ted did years of experiments, putting in many 24-hour days. I saw his unlined face and relaxed smile harden over the months, until the roguish hippie was nowhere to be found. But our results left us convinced that fly aging was pretty smooth: there were a lot of genes contributing to life span in our fruit flies. It wasn't lumpy. There was no "silver bullet" gene that determined when a fly dropped dead in our populations.

Drosophila is preeminently the "single gene" animal of genetic research, because we know so many individual fruit fly genes that have large effects. If the totality of aging in normal *Drosophila* is not due to a single gene, or a few such genes, there is little reason to expect that human aging is determined by one or two genes. The problem of aging is a problem produced by a monster with many heads.

But I was still interested in the question: how many genes control aging? And from that question, I wondered how many different biochemical processes normally control aging? This in turn will determine how hard it will be to postpone human aging.

I knew that I needed new research methods to address this question. The experiments that Ted spent five years of his life on were too crude. We needed more refined technology.

One of the interesting aspects of science in the late twentieth century was the proliferation of institutes funded by generous benefactors. Sometimes these institutes grew out of the initiative of very wealthy nonscientists who wanted to make a difference in the world. In other cases, leading scientific figures drew to themselves the resources to start an institute that expressed their interests. A stellar example of the latter is the Linus Pauling Institute of Science and Medicine, which was in Palo Alto, California, from 1973 to 1996. (It is now affiliated with Oregon State University.)

Linus Pauling started this institute out of frustration with universities. He wanted to do research on his vitamin C theory of health and aging. Pauling thought that the ingestion of large amounts of vitamin C along with other vitamins and nutrients was the key to the preservation of human health. He himself took two or more grams of vitamin C each day.

There were several points that promoted the acceptance of Pauling's beliefs. The first was that he was a Nobel laureate in chemistry, as well as being a member of the leading scientific societies, such as the American National Academy of Sciences. The second was that he himself lived a long, productive life on his regimen; he died in 1994, at the age of 93. The third was that Pauling was a gifted communicator, whose lectures and popular books attracted the interest of many people from outside universities and institutes. Pauling's preeminence and popular impact meant that he could attract the funds to pursue his maverick interests. Thus he was able to create the Linus Pauling Institute.

Jim Fleming worked at the Pauling Institute for quite a few years. He had already worked on the aging of fruit flies for a number of years when I met him in the late 1980s. He had never taken a PhD. Indeed, he didn't even get a bachelor's degree in biology until he had been involved in research for years. The Linus Pauling Institute was the perfect home for someone who had spent years ignoring the conventions of academia.

Jim was also unconventional in his choice of technologies. One of these was an apparatus that smears out many different proteins in the two dimensions of a thin slab of gel. Gels are a kind of scientific Jell-O. Biologists use gels to separate large molecules, like DNA and protein, from each other. Jim's 2-D gels gave a scatter-shot inventory of the proteins that had been recently synthesized. 2-D gels can display as many as several thousand proteins and most of these proteins will correspond to a particular gene. Therefore, 2-D gels provide a crude, but large-scale, survey of an organism's genetics.

I saw Jim give a seminar at the University of California–Irvine, shortly after I got there. He showed how he used big slabs of gel to survey the protein changes during aging in fruit flies. What was surprising to me then was how little the proteins of the fly changed with age. But I should have remembered a result that came out of Maynard Smith's lab in the 1960s: flies don't make much new protein as adults.

Jim had his hands on some great technology for the time. But he didn't really have a good biological problem to apply it to. I thought that I could give him the right problem, estimating the number of genes that control aging in my fruit flies. I had him mostly talked into it over dinner at a Chinese restaurant after his seminar. I followed that up with visits to the Linus Pauling Institute. By 1990, Jim and I were working closely together. I never spoke with Linus Pauling, though. I would see him at the other end of the parking lot getting into his beat-up car. But I hope that he approved of what Jim and I were doing.

Here's what we did. I drove from Irvine, in southern California, to Palo Alto, just south of San Francisco, with a box of living flies in their normal vials. I made sure that the flies were kept air-conditioned, so that they weren't under any stress. At the Linus Pauling Institute, extracts from each fly's tissues were placed in large 2-D gels that had strong electrical currents running through them. The charged proteins then migrated through the gel. Flies make more than 10,000 different proteins, but Jim was able to resolve only 321 proteins on his 2-D gels. Six proteins were strongly associated with postponed aging. This suggested that about 2 percent of the genes sampled by Jim Fleming were probably involved in the control of aging in our flies.

At that time the total number of normal genes in fruit flies was not known. Our assumption was that this number was likely to be between

10,000 to 20,000. (We now know that it is about 14,000.) If 2 percent of all loci control aging, we reasoned, then the total number of such controlling genes is between 200 and 400. This was more than enough genes for "smooth" inheritance. We published the paper showing this in 1993.

Even though we were happy with our estimate of the number of aging genes, we knew that our study was likely to be superseded soon by improvements in the molecular genetics of the fruit fly. And that is exactly what happened. I was amazed by the progress that would be achieved in less than a decade. By the turn of the millennium, the fruit fly genome had been completely sequenced, with almost all of the DNA information determined. Not only did the DNA sequencing of the entire fly genome give an excellent estimate of the total number of genes in a fruit fly, it also provided the specific sequences for all these genes.

With these gene sequences in hand, molecular biologists created *gene chips* with bits of DNA from each of the thousands of fly genes. Biologists use these gene chips to infer how much protein is made from a gene. When a lot is made, biologists say that a gene is "highly expressed." The key experiment for aging research was obvious: estimating the level of gene expression for all the genes in the fruit fly, as a function of age. Genes that have changing expression with age may be genes that affect aging. A colleague of mine at the University of Southern California, John Tower, performed experiments on gene expression in my flies, and the results were promising—it was indeed possible to show that some genes changed their gene expression with age, and a few of these genes were different with respect to gene expression in Methuselah flies, too.

The group that bit off the aging of the entire fly genome was at University College London. Their paper appeared in the journal *Current Biology* in 2002. They compared gene expression from the first week of adult life to about ten weeks of adult life—very late in life for a fruit fly. The data are a bit difficult to interpret. But a conservative estimate based on their data would be a gene number of 400 or 500.

It is interesting to compare this high-technology estimate with the crude one that Jim and I obtained about a decade earlier. If we multiply our estimate of 2 percent of the fly genome involved in aging by the correct genome size of about 14,000 genes, our estimate is about 280 genes controlling aging. This is gratifyingly close to the far superior estimates that can be derived from the data obtained by the University

College group. The correct number of genes controlling aging isn't two or five, and it isn't 10,000. It's a few hundred.

A similar research project with nematode worms gave an estimate of 164 genes showing a response to aging. Given the great disparities between lab fruit flies and lab nematodes, it is impressive that the number of genes implicated in aging, in some manner, is in the same ball park for the two species. We can be fairly sure that aging is not normally controlled by a very small number of genes, nor is it likely to be controlled by the majority of genes in an animal's genome.

At the end of this quest for the number of genes controlling aging, Leo's original result has not been supported. Still I think that the chill he gave me helped us to find our way to the truth. When a compelling idea like the many-headed monster of aging comes under challenge, the challenge motivates scientists to dig deeper. Like professional athletes facing elimination during the play-offs, you try harder.

Let's say that the correct number of genes affecting aging in fruit flies is approximately 300 to 350. Humans have at least 50 percent more genes than fruit flies. If everything were to scale up proportionately, then humans have about 500 genes that control their aging. With 500 such genes, no single one of them is going to supply a silver bullet that would alleviate all the ills of aging.

This implies that there is no master control for human aging. Biomedical research on the problem of aging will not come down to the search for a single "death gene." Unfortunately, such searches for simple master controls were the main strategy of twentieth-century medicine. It was a marvelous strategy for identifying pathogens and genetic diseases, even for treating diabetes, but it won't work for aging.

Yet none of that matters. We now have technologies that work with hundreds of genes at a time. A many-headed monster strategy for ameliorating aging is now at least possible. At the very moment when the aging field must come to terms with the intimidating magnitude of the problem of controlling aging we have technologies that can slay the many-headed monster.

13

Woody Allen and Superman

Many of my relatives have lived past the age of 90. But the one who always appears in my mind's eye as the epitome of longevity is my maternal grandfather, G. F. Horsey ("Fred"). He lived from 1880 to 1980, dying shortly after his hundredth birthday. There is something about that round number 100, about its stride across a century. My grandfather became an adult the year that Queen Victoria died, 1901. Yet I watched him see Richard Nixon get the 1968 presidential nomination on television, an invention that was barely imagined the year my grandfather was born.

I have a photograph of grandfather Fred from 1905. He is wearing a style of suit that only came back into fashion in the last few years: small lapels high on the chest, with three single-breasted buttons. He looks quite handsome, indeed manly. He was tall for his day, more than six feet in height, and broad shouldered. But he was always lean.

The family fortunes were not the best when my grandfather was a young man, even though his father, George, had been a successful Park Avenue dentist in New York City. George was a toothbrush inventor and entrepreneur, as well as a gambler, who spent faster than he earned. Though Fred started an engineering degree at Queen's University, Ontario, he had to quit to work in order to support his mother and his sister, Julia. For the next few decades, he built railroads and roads in the Canadian north and the Canadian Rockies. He was put in charge of crews of laborers hacking their way through tundra, swamps, mountain passes, and forests. He finally married at 50 years of age, and had two children. Eventually he worked as superintendent for Canada's national parks. He retired in 1945. A bit later Parks Canada named a Rockies mountain summit after him, a snowbound peak that looks like it has a nose, not far from Emerald Lake in British Columbia; the famous Burgess Shale is nearby.

When I knew my grandfather, he was a very quiet man, one of the few who read Hansard, the record of speeches in the Canadian House of Commons, the main legislative body for the country. He had dentures and occasionally smoked small cigars. He was very sentimental, and cried at family dinners. He maintained a regular fitness program into his nineties, running on the spot interminably.

In their long race toward destruction, his body outlasted his brain. My grandmother said that she knew her husband was in trouble when he suggested to her that they start saving for their old age. He was about 97 at the time.

In his last few years of life, he became confused about people. My mother visited him for his birthday and he confused her with his sister, who had the same name. But Fred still followed the World Series, which they played around the same time as his birthday, starting in 1903. They played more than 70 World Series in his lifetime. His doctors found some growths on his neck around his hundredth birthday, and they took him to a hospital to have them removed. Like many of the elderly exposed to the bacteria in hospitals, my grandfather caught pneumonia and died.

The scientists who work with centenarians are generally convinced that genetics plays a big role in whether or not you reach 100. If you have a sibling who reaches 100 years, you have a much greater chance of reaching 90 years, four or five times greater, according to the New England Centenarian Study. Children of centenarians also do well. The New England study is directed by Thomas Perls, a youngish man of great determination. The Perls group has also found a trend toward greater fertility after the age of 40 among female centenarians. Across national groups, centenarians are generally more robust, with a greater tendency to physical vigor and postponed dementia. Centenarians are not usually people who are institutionalized in their seventies, with a prolonged period of dependence.

The Perls group emphasizes that while there are some common features among centenarians, there are also significant differences in such lifestyle patterns as years of education, wealth, religion, ethnicity, vegetarianism, and exercise. Centenarians are by no means all ascetic or athletic. However, both obesity and smoking are very rare in the centenarians found by the New England team.

Whatever interpretation is placed on the causes of long human life spans, centenarians demonstrate what late life looks like. I think the most important thing about them is that they have a lot of *life*. Centenarians do not usually lead lives of great debility. They lead lives of great activity. In thinking of my grandfather carving roads and railway tracks out of the Canadian wilderness, any connection between long life and human invalidism is severed in my mind. One of the most important questions in the medicine of aging is the nature of the oldest humans, the centenarians. This question was to receive some remarkable answers in the 1990s.

Larry Mueller has been my most important colleague for the last 15 years. We both got our PhD's in 1979, working on evolution in *Drosophila*. We have known each other since 1983, and Larry was hired by the University of California–Irvine, in 1988, the year after I arrived there. Now our offices are next to each other. Larry and I have very different styles. He is quiet, even-toned, and methodical, while I am passionate and intuitive. Larry analyzes data, and I think up outlandish experiments. Larry reads the scientific literature, and I try to catch the scientific gossip. Larry is an athlete, with a running addiction that keeps him in great shape when his doctors don't restrain him from indulging his obsession. With his greater height and better looks, I sometimes feel like Paul Simon to his Art Garfunkel.

Larry came to me in 1992 with two papers just published in the journal *Science* that he knew I had yet to read. The principal authors of the two articles were James Carey, James Curtsinger, and James Vaupel, all well known to me. The gang of Jims. Carey was an entomologist, Curtsinger was a population geneticist, and Vaupel a demographer. The experiments they had published in *Science* used both medflies and fruit flies.

The papers claimed that there were no limits to life span in flies. Their findings were that aging stops, or even reverses, at late ages. In other words, their data seemed to show that a nonaging phase of life followed aging.

I refused to believe the results. I couldn't understand how aging might just stop, to say nothing of reverse itself. My view was that biological immortality was found only in organisms that reproduce by splitting symmetrically in two, like many bacteria or sea anemones.

It wasn't hard to find problems with the medfly experiment. The medflies had been kept in large cages each containing tens of thousands of flies. As the flies died off, they were kept at lower and lower densities. Joseph Graves, previously a postdoc in my lab, had shown that *Drosophila* longevity increased at lower densities. Larry Mueller had been involved in that work, too. Together with Ted Nusbaum, then my postdoc, the four of us wrote a note to *Science* arguing that the cessation of aging in the medfly results might be an artificial effect of falling density.

But we didn't have a way to get around the fruit fly results, because they were kept at lower densities, which reduced the density problem. Their aging still stopped. There had to be something there.

This problem nagged at me. Could aging really stop late in life? I tried not to think about it, without success.

After months of unease, I got a glimmer of an idea. We can explain aging using the force of natural selection. Once a population starts to reproduce, the force begins to fall. The force continues to fall until reproduction ceases. At that age, if not earlier, the force hits zero. This you already understand, if you have read this far. The interesting thing is that the force remains at zero for all later ages, once it first hits zero.

The important change in my thinking concerned the impact on death rates of this pattern in the force of natural selection. I had always thought that the zeroing of the force at late ages meant that death rates would reach 100 percent after the force of natural selection hit zero. But what if the increasing death rates of aging were produced specifically by the fall in the force of natural selection? And what if, once the force bottomed out at zero, the increase in death rates came to a stop? Since the force stays at zero at late ages forever, perhaps death rates would also plateau forever, starting very late in life? The death-rate plateau would be near zero survival, but it would be stable. Death would still occur, indeed often, but the implacably accelerating death rates of aging would be over.

The very old members of sexual populations would be like animals that reproduce by fission and do not suffer systematic increases in their death rates. The very old for whom aging has stopped would also resemble fissile animals in nature, such as sea anemones, in that they would not live very long. Most fissile animals are small and vulnerable. But they do not age. For this reason, they are conventionally called "biologically immortal," though there are many in the aging field who are squeamish about the term *immortal*.

I proposed to Larry that the late-life cessation of aging could be due to the end of the fall in the force of natural selection. He listened politely enough. We both realized that this idea had to be explicitly analyzed. I had a cheesy little mathematical argument for my theory, but it was too slight to deserve much credit. Larry dutifully set about creating simulation models of the evolution of aging, to see if my intuition was right. My hope was that Larry's simulations would produce populations that had a cessation of aging late in life at least some of the time. That was all I was looking for, because I thought my idea would only work in special cases.

By 1995, Larry had completed his work. I was astonished when every case that Larry modeled on his computer produced a late-life plateau in mortality rates. It didn't matter whether genes had effects over a few or many ages, whether there were genetic trade-offs or not, or what the population size was. Those things mattered only quantitatively. They changed how fast aging reduced survival, how soon the aging phase came to an end. But all of the simulations produced a two-phase adulthood: aging was followed by a period in which death rates either did not increase, or only increased at a much slower rate. The aging phase was just a transition between two relatively stable conditions, low juvenile mortality and high late-life mortality. These results were published in 1996 in the *Proceedings of the National Academy of Sciences, USA*.

I had several opportunities to present our theoretical study in workshops and small meetings on aging. Evolutionary biologists had some doubts, but they were generally interested. Mainstream aging researchers and demographers showed a mixture of incredulity and outrage. This convinced me that I had to be on to something good. Their reaction to my original work on aging had been just as hostile when I first presented it to them years earlier.

Part of the problem was that most people who thought about late life interpreted the mortality-rate plateau in terms of a theory called *demographic heterogeneity*. This theory has been developed primarily by James Vaupel, an American demographer who is now based primarily in Germany. He is the third Jim from the gang of Jims who destroyed the prevailing hypothesis that aging proceeds all the way to the end of life. For that, he will always have my respect.

Vaupel's theory makes very simple but very extreme assumptions. It begins by assuming that each cohort begins adulthood with a wide spectrum of *robustness*. (A cohort is a group having roughly the same age and who grow up together, like your classmates in grade three.) Some individuals are assumed to be robust, with a lower likelihood of dying their entire lives. Other individuals are assumed to be weak, with a higher likelihood of dying their entire lives. It is very important to notice that heterogeneity theory assumes that this heterogeneity of robustness is sustained *throughout life*. The robust stay robust. The weak remain infirm.

With these assumptions, it is inevitable that the weak will die off earlier. At later ages, the more robust will dominate among the survivors. At very late ages, only the very robust, with their very low death rates, are left. These are the individuals that survive the period of aging, making it into the late-life period of slower increases in mortality. A central feature of this theory is that it still assumes that all adults continue to age, so the late-life plateau in mortality rates is assumed to be merely a slow aging process. Aging never ceases under the assumptions of the heterogeneity theory.

This theory has to be correct *if* there is an enormous and sustained difference between the robust and the fragile members of each and every aging cohort. Whenever this precondition is true, heterogeneity will produce a demographic pattern of late-life slowing in the increase of death rates associated with aging. Given the assumptions of the theory, the expected result is inevitable.

But the requirements of this theory are extreme. The range of variation among the members of a cohort has to be vast. The theory requires a range in lifelong robustness like that between Woody Allen and Superman. That is, there have to be very weak individuals who die off early, so that at late ages the population consists overwhelmingly of vastly superior individuals who still age, but do so at a virtually indetectable rate. More importantly, the rank order of individuals has to be sustained. The star high-school football player would have to be the most robust 60-year-old at the class reunion, 42 years later. And so on for the ranking of all other individuals, from ordinary Joes and Janes, down to the nerds like myself. Yet, in real life, football players tend to die early. There is very little evidence for sustained Woody Allen–to–Superman variation in aging cohorts. Indeed, no one has ever found a single living cohort, occurring naturally, with enough demonstrably consistent heterogeneity to generate a substantial slowing of aging very late in life.

Yet such is the intuitive appeal of demographic heterogeneity theory that it remains, to this day, the explanation for late-life survival patterns that most biologists and demographers who think about late life assume is correct.

This annoys me. And when I'm annoyed, I want data. Larry and I proceeded in two directions: first, we tested demographic heterogeneity theory, and second, we tested our evolutionary theory of late life.

The problem with testing heterogeneity theories is that they are based on a hidden variable, robustness. What that robustness consists of, much less how to measure it, is unclear. Robustness could be due to anything, from stored fat to disease resistance. Most problematically, the unknown robustness variable gives the heterogeneity theorist almost unlimited opportunities to dodge experimental bullets. They can always claim that one experimentalist's robustness character isn't actually the robustness that their theory requires. That way, if an experimenter looks for the required lifelong heterogeneity in a particular character and fails to find it, the defenders of the heterogeneity theory can always claim that the test wasn't fair.

We thought we had a good opportunity to test heterogeneity theories because we had populations that varied in stress resistance *and* we had already shown that stress resistance determined life span in *Drosophila*, at least in part. Therefore it was reasonable for us to use stress resistance as an indicator of robustness in experimental tests of heterogeneity theory.

Mark Drapeau was working with Larry and me at the time, so we got him and a team of undergraduates to compare *Drosophila* populations that had considerably increased starvation resistance with populations that had normal starvation resistance. (Mark moved on to the molecular genetics of the *Drosophila* gene *yellow*. An ambitious young man, I very much regret that he did not remain in evolutionary biology. But the money is in molecular work, if not the science.) In other words, he compared more robust flies with less robust flies. The heterogeneity theory implies that the more robust should have reduced mortality rates in late life, postaging. They did not. The experiment did detect some large changes in mortality early in adult life, but no significant differences in late-life mortality. Mark's data didn't fit the heterogeneity model.

Larry had the next idea for a test of heterogeneity theory. He fit heterogeneity models to our laboratory's mortality data in *Drosophila*. He then calculated the predictions of the heterogeneity models that best fit our fly data. He did his calculations as if he actually believed in heterogeneity theory, to see what the consequences of that belief would be when he made assumptions that were the most favorable for heterogeneity theory. For example, he calculated the predicted age at death of the last fly to die in an experiment of our size, based on a heterogeneity model that had been modified to fit our actual data over the entire adult period as closely as possible. Intuitively, if the oldest individuals surviving in the population are vastly more robust, then the longest-surviving individuals should attain great ages before they die. When Larry did the actual number-crunching, it turned out that the best heterogeneity models predicted that the last fly to die in our experiments should have lived far longer than any real fly did. That is, the heterogeneity model made the wrong predictions, even when Larry deliberately fit that theory to our real-life data.

This happened because heterogeneity models have to assume a lot of feeble individuals and a lot of extremely robust individuals, the aforementioned Woody Allens and Supermen. Actual fruit flies don't exhibit such an abundance of extreme types. This was a second blow against the heterogeneity theory.

Our data haven't been unusual in giving heterogeneity theory a hard time. Jim Curtsinger, the person who discovered the end of aging in *Drosophila*, has also published some experiments that test heterogeneity theory. None of his results support the idea, and some provide evidence against it.

Things have now become interesting. No published evidence directly supports heterogeneity theory as of this writing. A growing number of experiments refute heterogeneity theory. But the theory still remains the first choice of a wide range of gerontologists and demographers. I will not now name them so they can deny their adherence to heterogeneity theory later, like the Vicar of Bray who switched between Catholicism and the Church of England as one Tudor monarch succeeded another. The only true believer in the theory, who would assuredly choose execution over expediency, is James Vaupel, its primary architect. At least one has to admire his tenacity. Science needs more people with his kind of courage. Too many scientists just sail with the prevailing winds, like the Vicar of Bray.

As you know, there is an alternative to heterogeneity theory, the evolutionary theory of late life. It has received some criticisms from Jim Curtsinger, Scott Pletcher, and Ken Wachter, among others. Many of these criticisms are valid and point to the need to improve on our first crude models. That's the way theory usually develops. Recently Brian Charlesworth has supplied a powerful treatment of the evolution of late life, far superior to anything else in the field. He finds plateaus like the ones Larry found in his simulations. With people like Brian working on it, I think that late life will soon have an evolutionary theory of great power.

For me, however, the most important test of a formal scientific theory is not whether theoreticians and applied mathematicians can find a way to improve on it. I expect them to do so, because that's how they pay for their Cheerios. Rather, the better test of a theory is its success in surviving challenging experiments, experiments that put the theory at risk of annihilation.

So that's what we've done to *our* theory of late life, based on evolution. Our first experiment was a variant on the selection procedure used to create our Methuselah flies. In the evolutionary theory of late life, the age at which reproduction *ceases* plays a similar role to the age of first reproduction in the evolution of aging. As that *last age of reproduction* goes up and down, aging should cease later and earlier, respectively. Larry has shown that evolutionary theory generates this prediction using explicit calculations; we aren't just guessing about this. The plateau in mortality rates occurs only *after* the force of natural selection hits zero, which is about when reproduction stops.

The experimental challenge was clear. In flies that have had many generations with different last ages of reproduction, the age at which aging stops should follow the same pattern as the last age of reproduction, increasing when that last age is increased for many generations, decreasing when it is decreased. We tested this in 25 different *Drosophila* populations, with a range of ages at last reproduction from 9 days to 70 days. The late-life data showed that the age at which aging stopped did indeed rise and fall as the last age of reproduction imposed on the lab populations for many generations rose and fell. This pattern fit the predictions of evolutionary theory. This was also the first time that *any* theory of late life had received direct experimental support.

But I wanted to go one step farther. The heterogeneity theory concerns mortality rates. With lifelong heterogeneity in robustness, some effect on late-life survival is expected, though this effect is probably too small to explain the virtual cessation of aging late in life. Our evolutionary theory of late life is an alternative to heterogeneity where mortality is concerned. But our theory of late-life mortality comes associated with theory concerning the evolution of fecundity. Fecundity has its own force of natural selection, separate from the mortality force.

There is no reason inherent to the heterogeneity theory why fecundity should exhibit a late-life plateau without further decline. The heterogeneity theory is all about the loss of individuals with low robustness, which is a qualitatively reasonable idea even though it doesn't work quantitatively when you look closely at actual data. But it supplies no warrant for loss of individuals according to their fecundity. Such a theory might be invented. Demographers are ingenious at inventing theories. But our evolutionary theory of late life *requires* that there be a late-life plateau for fecundity, because its force of natural selection also goes to zero and stays there forever after, just like the force of natural selection does for mortality.

My graduate student Casandra Rauser—think *Saturday Night Live*'s Tina Fey—went looking for evidence of a late-life plateau in fecundity among fruit flies. Given enough flies, it isn't hard to find. She now has evidence for the existence of a long flat tail in *Drosophila* fecundity at late ages. That is, fruit fly fecundity has a postaging late life, as evolutionary theory requires.

Heterogeneity theory remains highly popular among scientists. Unfortunately for them, it is entirely bereft of support from experiments that critically test the theory. That can be a problem, at least in the long run.

On the other hand, the evolutionary theory of late life has received significant experimental support. The evolutionary theory of late life is now undergoing refinement, both mathematically and experimentally. As time goes by, I predict that it will become harder for the proponents of lifelong heterogeneity to ignore or disparage the evolutionary biology of late life. The original evolutionary biology of aging, a somewhat different topic, has already received hazing from the scientific aging fraternity. First it was seen as absurd. Then it was considered just patently wrong. Now it is dismissed as obvious or trivial, so that it can be simply overlooked by those who choose to ignore its role as the scientific foundation for all research on aging. I am confident that the evolutionary biology of late life will proceed through the same stages of appreciation in the scientific community.

Let's talk about humans now. Over 95 years of age, human mortality rates stop increasing exponentially with age. Our death rates are phenomenally high in our nineties. But the only reason so many people survive into their second century is that there is an end to the rampant acceleration in death rates that occurs from age 15 to 90.

The small number of humans that make it to 100 might suggest to you that late life doesn't really matter to the postponement or retardation of human aging. Let me explain why it does.

Most scientists who studied aging before 1990 thought of aging as a rising wall of death, accelerating steeply to 100 percent mortality. I certainly did. We thought of the problem of postponing human aging as the problem of how to move this wall of death to later ages.

But we were all wrong. Aging is *not* an infinitely high wall of mortality, rising faster and faster as we get older, until everybody is dead. It is a ramp that takes us from a phase of low childhood mortality to a much later phase of high, but relatively stable, mortality. Postponing, retarding, or otherwise mitigating aging does not require pushing back a wall of death of infinite height. It requires smoothing out a ramp of mortality, and possibly lowering the height of the top of the ramp. Death is not quite as implacable as we supposed. This warrants hope where the substantial postponement of human aging is concerned. Perhaps enough hope for us to do something about it.

14

Not Even Oppenheimer

The evolutionary biology of aging leapt ahead in the last third of the twentieth century, first in theory, then in experiment. We now have answers to the most fundamental questions about aging. We can readily use evolutionary techniques to postpone aging in experimental animals. If the history of physics is a useful guide, and scientists tend to look to that history for guidance, we should now be able to build the gerontological equivalent of an atomic bomb: substantially postponed human aging. That is, enable people to play good tennis at 150 years of age.

Several technologies had to be improved to build Oppenheimer's atomic bomb: metallurgy, isotope separation, and electrically synchronized detonation among them. Likewise, additional biotechnologies are required to postpone human aging: whole-genome sequencing, high-density microarrays for gene-expression assays, genetic engineering, stem-cell culture, etc. Fortuitously, all of these technologies are being improved rapidly by cell and molecular biologists. Technology is what they are good at. These new technologies are much more powerful now than most biologists thought possible in 1990. I, for one, never imagined that biology would have such tools in my professional life.

Yet very little has been accomplished toward truly postponing human aging. It might be more accurate to say that *nothing* has been accomplished.

Important achievements do not come out of thin air. They are produced by exceptional people who take advantage of propitious circumstances. There are two obvious role models for the organization of great feats of technology. The first is Robert Oppenheimer's Manhattan Project. Oppenheimer was not necessarily the most creative physicist of his day. He wasn't even the person who first thought up the strategy for building an atomic bomb. That was Leo Szilard. But Oppenheimer brought the Manhattan Project to completion, overcoming the bureaucrats in Washington and the scientists he had to corral, exhort, and lead.

The second role model is Thomas Alva Edison's electricity R&D, the first large-scale R&D enterprise. Edison took the science of electricity, as developed particularly by Michael Faraday, and made from it a cornucopia of electrical inventions: improved telegraphy, the telephone, the phonograph, the incandescent light bulb, and so on. Edison wasn't always first with his inventions, the telephone being an example where another, Alexander Graham Bell, came first. Edison's inventions weren't even necessarily the best of their kind. But his company cranked out so *many* electrical inventions that electrification became a backbone for industrial civilization, instead of the marginal contributor that it was before Edison turned to it.

While the treatment of human aging needs its first successful technologies to get going, these technologies won't get the commercial aging industry off the ground without a singular leader pulling together the resources and talent necessary for large-scale development. This isn't easy. For about a decade, conditions have been ripe. The science and technology have been available. Even capital was abundant in the 1990s.

While my experience is by no means typical, I have spent much of the last two decades trying to develop treatments of human aging that will actually be effective. Since I have had no success as of this writing, my experience furnishes a curriculum in what the Oppenheimers and Edisons of aging face. I wish to outline this curriculum here.

In 1984, *New Scientist* asked me to write a major article on aging for their magazine. I duly wrote up a summary of research on the evolutionary biology of aging, but did not include any discussion of postponing human aging. There was a good reason for this. I had never thought about the problem of postponing the aging of my own species. I just postponed aging in fruit flies.

How could I have been so obtuse? Wasn't aging of preeminent medical importance? I had a few excuses, none of them impressive. I was still in my twenties, so I was emotionally disconnected from aging. Also, evolutionary biology at that time had very little interest in medical practice or pharmaceutical development. Professors are rewarded for their scholarly productivity, which has nothing to do with efforts to save mankind, the world, or, especially, trees. Our efforts are weighed for their paper and ink, and then turned into promotion or dismissal. So I had to be slapped around by *New Scientist* before I would pay attention to the problem of postponing or slowing human aging.

But they did wake me up, and I wrote a few paragraphs on the problem of postponing human aging. I proposed that we select on mice as I had selected on fruit flies, by delaying their reproduction, in order to make Methuselah mice. The Methuselah mice would enable us to determine the pathways that could slow human aging. I think it was one of the more timorous proposals for postponing human aging that has ever been published. But it placated my editors. Naturally they got rid of my dull academic title for the article. They changed it to "The evolutionary route to Methuselah."

They made the article the cover story. It was then picked up by the wire services, especially the idea that postponing aging has something to do with delaying reproduction. Within hours wire service articles had spread a very distorted version of my ideas around the world. My office phone started ringing in 1984, and it has never stopped. ABC, BBC, CBC, newspapers, magazines, radio, they've all asked for a sound bite or interview. Letters arrived from people grateful that I had endorsed their lifelong celibacy. That was kind of embarrassing.

My misadventures picked up speed in the fall of 1988, during a scientific meeting near Bar Harbor, Maine. Bar Harbor is one of those charming New England tourist traps with a tiny downtown, the kind of downtown where you can start an evening in a toney restaurant and proceed by foot through a succession of watering holes. I was familiar with this lifestyle from my years in Halifax, where Nova Scotians and visiting sailors drank like it was religion without having to get into cars and kill people from behind the wheel. Bar Harbor was also host to the Mecca of mouse genetics, the Jackson Laboratory, located a few miles outside town. David Harrison of the Jackson Lab was hosting the meeting, which concerned the genetics of aging.

Brian Charlesworth and I spoke on the first day, separately. Toward the end of my talk, inspired by the Jackson Lab's mouse facility, I brought up my suggestion to select on mice for postponed aging. I still wasn't enthusiastic about proceeding from flies to humans, but I felt that longer-lived mice would be ideal as a starting point for postponing human aging.

A day or so later I was talking with Huber Warner on the sunlit veranda of a cabin in Bar Harbor. Huber was my grant supervisor at the National Institute on Aging. He deserves credit as the person who got the NIA to fund good genetic research on aging. There was a rocking chair and the sunlight was quite relaxing.

Huber asked me, "Can you guarantee that your plan to breed mice will be successful?"

Guarantee is a word that I don't like in experimental science. I hedged. If you start with genetically variable mice, I said, and you use a lot of mice in your breeding program . . . then, yes, I could guarantee the creation of longer-lived mice.

They would be a wonderful resource, for a lot of researchers, opined Huber.

I concurred. This would be a great project for the NIA. You could foster aging research and the creation of treatments to postpone human aging, I said to Huber. It would be a groundbreaking government science project, like the Manhattan Project that built the atomic bomb. I already saw the need for an Oppenheimer.

Huber had a hungry look. It made me hopeful.

At the meeting, I was staying with the other participants in a mansion that had been deeded to the Jackson mouse lab. It was isolated from the road into town by a very long driveway that made its way through thick woods. The building was sited on a hill overlooking the sea, sharp rocks waiting to slash feet where the hill met the water.

At night, most of my fellow scientists played poker in their underwear, smoking cigars and lying about their sexual experiences. A fair bit of money changed hands, but there were no fistfights. I of course did not play cards, or smoke cigars.

I was bemused by a large painting of a beautiful young woman in a ball gown, which dominated the entryway. I was told that she had died on the *Titanic* coming over from England to marry the owner of the mansion. He had never married, living alone, becoming progressively more melancholy through the dark stormy nights of Maine, the waves crashing onto the sharp rocks below. I would have been more impressed by this story if I hadn't heard it before about a mansion in Halifax. But it is possible that both stories were true. Halifax was a common port of call for the transatlantic trade when the *Titanic* sailed.

Toward the end of the meeting, there seemed to be a consensus that my mouse-breeding project had some merit. Huber said that he would get in touch with me once he had checked with his superiors.

You might think that this meeting would have been an excellent opportunity for Brian Charlesworth and me to pat each other on the back over the burgeoning influence of our findings on American science. But neither of us is like that. Instead we played a board game with Joe Graves, then my new postdoc, and a few others on the last night of the conference. It was a windy night at the isolated mansion, as we pursued our juvenile amusement. Few of the other participants remained.

There was a sharp bang. The lights flickered. Joe cleared his throat, but we went on playing our board game. Brian was winning, and he was especially gleeful.

The wind really picked up, and there were more bangs.

People must have left their doors ajar when they left, I commented.

Brian just ignored the distractions.

Joe wasn't doing well. He paced around the table talking about horror movies. He wondered if the bereaved owner was haunting the mansion.

A rapid succession of bangs decided things for Joe. He yelled, "I'm getting out of here," then fled the building out into the darkness of its endless driveway. It would be a long run back into town for him, but I knew he had the adrenalin to go the distance.

By then it was obvious that Brian was going to win the game. We all went to bed.

When I woke the next morning, I was the only guest left in the house. The day staff told me that everyone else had left because of the banging. Possibly the mansion was haunted. Or maybe the Northeast winds make old houses creaky.

Back in Bethesda, Maryland, home of the National Institute on Aging, Huber managed to come up with some money for a workshop to evaluate my mouse idea. I was told to find an appropriate panel and arrange a meeting place for 1989.

I invited some of my heroes and some of my best colleagues. We met in the Beckman Center of the National Academy of Sciences, next to my University of California campus, in Irvine. The National Academy building could be described as modernist hacienda, but it is a great site for a meeting. Its complete artificiality clears the mind.

What I didn't realize then was that the start of this meeting marked the high-water point for my mouse project. While some of those at the workshop seemed to favor my idea, George C. Williams being one, virtually everyone had an alternative plan that they thought the NIA should fund instead. Tom Johnson, the founder of nematode-aging genetics, felt that mutation hadn't been tried enough as a method for postponing mouse aging. He was also skeptical about the prospects for analyzing the results of the selection experiment that I proposed.

There were several more meetings over the next few years, but as the meetings went by, my mouse-selection idea became one of several alternatives. Then it became a minor proposal for further consideration. Finally, it was eliminated from the discussions. The panels originally assembled to evaluate my plan became a forum for NIA grantees to talk with each other about their research. In the end, I wasn't invited to attend anymore. I was no Oppenheimer. That much was clear to me.

No other plan took the place of the one I proposed. The NIA just went back to doling out money to miscellaneous grantees. This is an approach that works well scientifically, where a diversity of ideas and experiments is important. But it is the opposite of making a practical difference to the achievement of a concrete objective.

Both Oppenheimer's Manhattan Project and the Human Genome Project required the marshalling of considerable resources to single ends. The Manhattan Project was secret, but the Human Genome Project was not. Large quantities of ink and paper were expended debating its feasibility, its advisability, and its theological significance. Some people just love to read their words in print. I can understand; I'm that way myself.

But the concerns of scientists were different from those of the hand-wringing ethicists, journalists, and politicians. When the sequencing of the complete human genome was first proposed, the concern of many scientists was that it would gobble up resources that otherwise might go to their own research. This criticism was addressed by the federal government by diffusing the sequencing work over many labs.

The plot then had an interesting twist. Craig Venter publicly proposed to sequence the entire human genome using the resources of his company, Celera Genomics, before the government project could finish. The government had to jettison its dispersed approach and concentrate its DNA-sequencing resources in fewer hands to get the job done fast enough so that Celera wouldn't beat it to the goal.

The lesson to be learned from the Human Genome Project is that big government research projects have to be steadily focused on their practical goals, or the infighting for money will diffuse, or derail, the project. I failed to control NIA's focus. My colleagues hijacked the initial interest in my mouse-breeding project, using it instead to advance their own specific projects.

An irony of this outcome was that I never had any interest in becoming a mouse breeder myself. My scientific interests were better pursued using fruit flies. But at that time I felt that the fastest way to advance toward the goal of postponing human aging was to create longer-lived mice. I was trying to initiate a project without a clear understanding of the institutional difficulties. Perhaps I should have sought a powerful NIA staff person to take up the project for themselves, but I didn't think of that at the time. I hadn't read enough Machiavelli.

Within a few years I had mostly lost sight of my plan for a government mouse-breeding project, or indeed any kind of Manhattan Project on aging. I was more interested in working with the faculty who had come together at UC–Irvine to study aging in fruit flies: Joe Graves, Tim Bradley, Larry Muller, and Allen Gibbs. Francisco J. Ayala, a National Medal of Science winner and leading evolutionary biologist, also helped. Along with us was a group of postdocs, graduate students, technicians, and undergraduates that would number more than 200 at its peak. This work was supported by both the NIA and the National Science Foundation.

We also received private support from the Tyler family in honor of Robert H. Tyler, a graduate student of Francisco Ayala's, who died while he was completing work on the molecular evolution of aging. Robert was awarded his doctoral degree posthumously. He was a very impressive human being with a wide range of talents, from science to piano and performance art, and I miss him to this day.

But the idea of postponing human aging wouldn't leave me alone. In 1991 and 1992, I gave a variety of interviews on the subject to popular magazines, even business magazines. Out of the blue, I received a request for a business plan from a venture capital firm, Menlo Ventures. I was flummoxed. I had no idea what a business plan looked like. I didn't know then that writing a business plan was the major extracurricular activity of the 1990s, safer than adultery, paper-cuts being the main health risk.

I learned a bit about writing business plans by attending seminars and buying software. With Ted Nusbaum, Robin Bush, and a few others, I put together a business plan that centered on the creation of mice with postponed aging using selection, as I had first proposed in 1984 in *New Scientist*. We called the company Methuselah Research. Off went the business plan to the venture capital firm, and we all had visions of spending our first million. I had fantasies about becoming the Edison of aging.

It wasn't quite crushing when the VC rejected the plan without comment, but I can't say that it stimulated a desire to write another business plan.

More time went by. I stuck with fruit fly research.

In the winter of 1993 I went to Lake Tahoe for another aging conference. This one was different. It was organized primarily to foster skiing among its participants. Going off the side of a ski-jump years earlier had left me unable to participate in the Alpine pleasures. Instead, I spent my spare time talking with people about how to postpone human aging, mostly for fun.

This wasn't entirely at my initiative. There was a crude pamphlet at the meeting with a cover that said, "Aging sucks!" It caused quite a stir among the assembled aging scientists, who pursued their trade with hardly a thought of changing the life spans of the people around them. To raise the subject of doing something practical about aging was considered bad taste for many of my middle-aged buddies. Ten years earlier I would have agreed with them.

Then I met the author of the pamphlet in a hallway: Ralph Andrews. Ralph was a television producer, mostly of game shows. He had a completely shaved head that made him weirdly ageless. In his sixties, he urgently wanted a solution to the aging problem.

Accustomed to applying personal pressure to get his projects funded, Ralph was trying to push aging scientists and NIH bureaucrats to do something. This was impressive. Most people who are like Ralph Andrews give money to quack herbalists or glib MDs, from whom they get nothing more than placebos and reassurance, if they are lucky. The unlucky have their health compromised.

I ended up at a dinner party that Ralph put on at a restaurant. The food was overpriced, the NIA bureaucrats were speaking in bafflegab, and my fellow scientists were terrified of offending somebody. But to me postponing human aging was a straightforward project. I said so, but wasn't given the chance to elaborate at the time.

Exciting things happen at conferences only late at night. Toward the end of a long day, I ran into Parviz Sabour, who worked with some postponed-aging mice that had been created at the Experimental Farm, in Ottawa, Ontario, a few miles from where I had attended high school decades earlier. The experiments weren't large, but Sabour's colleague Jiro Nagai had reasonable evidence that they had actually postponed mouse aging by the method I had proposed, delayed breeding. The scientists in Ottawa knew my work, so Parviz and I hit it off immediately.

Then Parviz and I ran into Bill Andrews, Ralph's son. Bill was a biotech scientist. Clearly, he was the one who had guided Ralph away from quacks, toward science. The three of us had a great time fantasizing about how to postpone human aging, using bred mice, molecular technology, and genetic analysis.

As our enthusiasm rose to fever pitch, Bill challenged me over my plans. "Why don't you start an anti-aging company?"

First I explained that I had already tried, once, but hadn't been encouraged by the response. Then I returned volleys. "I was thinking that you should start a company. You have industrial experience. I only know universities."

Bill said that he couldn't do it without backup from people with aging-research backgrounds. I offered to help him, and Parviz chimed in too.

We ended up around a kitchen table, planning to form a company to breed mice for postponed aging on a large scale, with Bill Andrews as the CEO. It was heady. This might be the full Edison, I thought.

By the next morning everything had changed. A heavy layer of snow had fallen on the resort during the night, and ice covered the walkways between the resort's alpine buildings. Bill summoned Parviz and me to the chalet where his father was staying. The walkways were treacherous and I could see my breath in the air for the first time in months. I had started to become a real Southern Californian. At the chalet meeting, Ralph told us that he wanted to be on board, in the biggest way. Bill explained that he felt his father could help with raising funds and other executive tasks, things Bill didn't think he was as good at.

Eventually Parviz and I agreed. What happened next was a whirlwind that would last five years.

The partnership that started that day didn't last long. Bill backed out because of his employer. Parviz had problems with the Canadian government. Ralph brought other people in. I brought Ted Nusbaum in. Some of Ralph's people left. Ted left.

After some months, the company was incorporated as MRX Biosciences. The kernel of the company strategy came from the Methuselah Research business plan, only Ralph made it more ambitious. I had told him that the bigger the selection scheme, the faster useful results would come. Ralph propelled everything forward like a man with a demon on his shoulders. He combined drive, focus, and a really ineffable charm.

But the company did not prosper. After years of frustration, I quit my post as consultant to MRX Biosciences. Extensively breeding mice for postponed aging would be a valuable approach to developing tools for slowing human aging. But I doubt that anyone will try it until the business of treating aging develops a large capital base. Unfortunately that in turn requires some successful aging interventions, which means that at least some aging-amelioration technologies have to be successful first. It is a situation somewhat like Yossarian's Catch-22: you can't get out of the army unless you're insane, but wanting to get out of the army shows that you are sane. We can't get reasonable funding to treat human aging until we have already succeeded at slowing or postponing human aging.

Giving up provides an opportunity for new ideas. In 1999, as I endured yet another droning colleague at yet another aging meeting, I began to daydream about the new genomic technologies: the completely sequenced genomes, the assays of the expression of every gene in the genome, the parallel work that had been done between fruit flies and humans.

Then it hit me. The genomic bridges between fruit flies and humans were strong enough by 1999. We didn't need mice for the R&D. We could go from flies to humans in order to find aging genes, even aging therapies. Mice would be good for pharmaceutical testing, but we didn't have to breed mice for five or ten years first. I was re-energized.

Soon there was another business plan and a new group of people to work with, Tony Long, Larry Mueller, Greg Stock, Pete Donald, Linda Howell, among others. Some came, some went. Again business presentations were made. Handshakes took place. At one point there was a verbal commitment of 12 million dollars from one investor. Then another investor committed to 8 million.

We never cashed a check. It all disappeared in the stock collapse of 2000, especially the collapse of the NASDAQ. It had been a narrow window of opportunity. The technologies we needed weren't all there until 1999, just months before the required capital started to melt down. If I had thought up the fly-to-human plan six months earlier, it might have been funded. But I missed my slender opportunity.

There are bound to be serendipitous discoveries of therapies for human aging. And one of these could be the therapy that leads to the creation of an industry that transforms human aging. I am confident that this industry will eventually be successful. If I had acted earlier in 1999, there might be medications in clinical trials now.

The amelioration of aging is an achievement that is still being resisted. Most people do not believe in it, and members of the medical and scientific establishments are generally opposed to it. The June 2002, issue of *Scientific American* contains an article entitled "No truth to the fountain of youth," by Olshansky, Hayflick, and Carnes. Linked to this article is a longer statement of similar import signed by 51 scientists. The message is simple: there are no proven therapies that ameliorate aging. This ineluctable conclusion is qualified to the extent that Olshansky et al. admit that further research might someday yield such amelioration of aging. But they don't sound very optimistic. For members of the anti-aging movement, this *Scientific American* article is an emblem of their frustration, a cold, hard, bitter pill to swallow.

In his book *The Tipping Point*, Malcolm Gladwell shows how ideas that are dormant at first eventually start to spread. He finds that these ideas are not necessarily improved as to content. Instead, small quirks of their presentation take the previously neglected and make it into something of interest to a great many people.

How will the treatment of our aging achieve its tipping point? How will it go from fervent hope to medical practice? Many of the examples in Gladwell's book concern marketing and media. He discusses how Hush Puppies became fashionable and why *Blue's Clues* is a hit with preschoolers. In these examples, small details of presentation made all the difference: repetition, gimmicks, context. If one were to apply these lessons to the alleviation of aging, new slogans, advertising campaigns, or spokespeople might well do the trick.

But if there is anything that the treatment of aging has received over its multimillenium history of failure, it is ingenious packaging. For every hostile Olshansky or Hayflick, there have been dozens of silver-tongued hustlers more than able to convince customers or followers of the value of an anti-aging nostrum. Taoism erected an entire cosmology with the treatment of aging as one of its central themes. This cosmology still influences millions in Asia and some thousands in the Western world. Despite Malcolm Gladwell's fascinating examples of the spread of ideas and products, his favorite devices are unlikely to improve the treatment of human aging.

Perhaps another parallel offers more promise. In the rampant venture capitalism of the 1990s, entrepreneurs waving business plans sometimes got funded if their plans were sufficiently eloquent, especially when those plans were to be played out on the Internet. But the thing that best guaranteed funding from a VC firm was a killer application. The query "Where's the killer app?" was a leitmotif of business-formation meetings. In growing the business, the company's team would seek the first killer app.

Examples of this pattern are not hard to find. Personal computers were a hobbyist's game until the Apple II transformed word processing in 1980. Then many people who weren't interested in personal computing, as such, had a good reason to go out and buy an Apple to increase their productivity at the desktop: the word-processing killer app.

So long as automobiles were unreliable contraptions that broke down every few miles, and cost far too much, few people wanted them. Once Ford put affordable, fairly reliable Model T's on the market, the automotive industry took off. The Model T was the killer app.

Heavier-than-air flight was disparaged by the professors and engineers of the nineteenth century. After the Wright brothers flew the *Kitty Hawk*, the airplane became one of the most important new technologies of the twentieth century. All the naysayers who came before the Wright brothers are now forgotten.

So long as portable phones were expensive, car-bound devices, only Wall Street and Hollywood moguls bought them. But the inexpensive, small, convenient cell phones of the late 1990s became hugely popular, the killer app of the mobile communications revolution.

I think the situation facing the treatment of human aging is simple. The anti-aging therapies that are now on the market have little clinical support, if any. There is nothing that indubitably works. This predicament is not helped by the unsupported claims of some companies and clinics that they have genuine anti-aging products. In the case of, for example, skin-care products, the amelioration of aging is inherently inconsequential. No one is going to die a horrible death from wrinkles or dry skin. Likewise, the vague claims made on behalf of herbal preparations and exercise programs hardly inspire much enthusiasm. Many treatments in modern medicine and sports training can achieve more. Demi-monde anti-aging medicine is an aggregate of distractions for the overly hopeful, not a revolution in human health.

The treatment of human aging is waiting for the first killer application of gerontological science. It needs a medication that will significantly slow the progression of life-threatening pathology. This pathology might be cardiovascular disease, cerebrovascular disease, cancer, or Alzheimer's disease. It doesn't matter which. It is important that this medication not be a mere treatment that alleviates or remedies a pathology once it has arisen. That is conventional medicine. True amelioration of aging must intervene in predisposing processes, processes that will produce disease and debility during the course of aging.

Let's be concrete. Suppose a scientist who works on cell biology finds a water-soluble chemical that prevents human cancer cells from violating the Hayflick Limit. After a number of cell divisions, cancer cells would be unable to continue reproducing, terminating tumor growth. All tumor growth. If we took two of these pills everyday, we might never get cancer. One of our greatest scourges would be neutralized. This medication probably wouldn't be perfect. People under 40 might have to forego it in order to maintain their fertility—making sperm and growing fetuses involve considerable cell division. There might be some side-effects on the immune system, because of the extensive cell replication that occurs in the production of antibodies. Yet the virtual elimination of cancer would be a medical benefit of staggering proportions.

The interesting thing would be the next step. Once people realize that modern biological research can forestall one of the diseases of aging, there will be a great hue and cry for similar work on other diseases. One by one, the diseases and pathologies of aging would be conquered. As this unfolds, people will live longer and longer, with better and better health. New degenerative diseases will become common, as people survive the ones that used to kill them at earlier ages, and their remedy will be the business of new generations of medical scientists. The treatment of aging will by that time be one of the most important tasks of physicians and one of the major targets of pharmaceutical development.

There is nothing unusual about this cascade pattern. In the nineteenth century, it was the pattern of the railroad and the steam engine. First a few railways were built, and regarded as curiosities. Then enough were built to be useful, and soon everybody was laying track and building steam-engine trains. In the twentieth century, this pattern was followed by the automotive, air transportation, and electronics industries, one after the other. In the twenty-first century, I am confident that the treatment of aging, among other applications of biological research, will also follow this pattern.

But it all has to start from a killer app: one case in which a new aging therapy is used successfully. This success must be practical. When

I started work on aging in the 1970s, it was sometimes said that the substantial postponement of aging in an experimental animal would transform the world. Having been the first to accomplish that deliberate postponement myself, I know now that this statement was not true. Most people don't care how long fruit flies live. We can double or triple their life spans, and the world won't change. Practical success is what matters. Practical success in the treatment of aging means demonstrably postponing or retarding human aging. This would be the flight of the *Kitty Hawk* for the treatment of human aging.

There are those in the anti-aging community who contend that they have already found a killer app: human growth hormone. Supplying middle-aged and older patients with growth-hormone injections leads to a spectrum of effects that reverse symptoms of aging. Muscle mass increases. Fatty tissue decreases. Muscle strength and overall energy increase. Many claim improved male sexual potency, though data on normal male potency are extremely hard to come by. Numerous patients are very happy with the effects of growth-hormone treatment.

On the other hand, growth-hormone treatment leads to a variety of minor health problems. Generalized edema occurs. Some male patients experience breast development. Carpal tunnel syndrome develops in some cases. Chronic pain afflicts others. In a few cases, growth-hormone treatment has even led to diabetes.

Of greater concern is the potential for growth hormone to foster cancer. Growth hormone might foster the growth of tumors that are already present. This presents a particularly worrisome scenario, because older patients might have small, occult tumors. Since you do not have to be a clinical oncology patient to have tumors over the age of 60, growth hormone might convert small, slow-growing tumors, which don't normally kill the elderly, into lethal growths. Direct evidence for this cancer-risk scenario is fortunately limited at present, but it worries me nonetheless.

It may be of interest that mutant mice that have little or no growth hormone live substantially longer than normal mice. However, mice of this type generally have low to nonexistent fertility, which we know is likely to increase longevity in itself. Despite this, the basic research on growth hormone suggests that supplying more of it is inimical to later health, not beneficial. There are admittedly many complications to this, or any other, interpretation of the effects of growth hormone. However, it is a reasonable summary to say that it has not been shown that the human use of supplemental growth hormone is an effective treatment for aging.

Testosterone is another hormone that is sometimes given to older patients. While it is not known to foster tumor formation, it does promote the growth of some tumors when they are already present. Growth hormone and testosterone treat fairly minor age-related conditions, especially the weakening of the aging male. Hormone supplementation does so at some risk of both well-established minor health problems and less certain major health problems. No hormone supplement has been shown to eliminate any major mortality risk in normal patients, leaving aside the use of insulin for diabetics.

Despite the sense of satisfaction that some patients experience under hormone treatment, hormone supplementation is hardly the killer application that is needed by the nascent aging industry. The known benefits are relatively minor, while the known and the conjectural health risks are significant.

The need for caution is obvious from the recent scandal over HRT, hormone replacement therapy. HRT is mainly the provision of estrogen and related hormones to postclimacteric women, whose ovaries have largely shut down hormone production. Despite many beneficial effects from HRT, which range from reduced osteoporosis to increased sex drive to relief from mood swings, the net impact of HRT on survival is negative. Women who take estrogen supplements long after menopause may run an increased risk of death from cancer and other diseases, at least in some studies, without a fully compensating reduction in other risks of death. These findings have arisen only recently, despite the long history of HRT, and many clinical studies of its medical effects. Growth hormone is much less characterized from a safety standpoint, compared to HRT, and there are better arguments for it being a medical liability. After the HRT story, we can add that this risk from growth hormone applies *even when* reduced levels of growth hormone are found clinically. All HRT patients have such reduced levels of estrogen, but that does not make HRT safe for them, on average.

Some day hormonal treatments will be an important part of the treatment of aging. There are many hormones that act to regulate human physiology, with effects on both cancer and vascular disease. Substantial progress in the treatment of aging will require the delimitation of hormonal influences on aging. This is a daunting task because hormones are, by their very nature, caught up in interactions with many body functions and many other hormones. They are the signaling and coordinating agents of the body's physiology. This gives them great potential as agents of aging therapy, but at the same time this makes them scientifically challenging. For now, we simply do not know enough about hormones.

If hormones are not likely to supply us with the first killer application, where and how can we look for that application? I see two possibilities, which are not mutually exclusive. The first is that we go ahead and implement the kind of genomic strategy described here, starting with model organisms and proceeding through the human genome to pharmaceutical development. I expect medicines of some value to come out of this program.

The second way of looking for a killer application is far more opportunistic. Vast numbers of legitimate pharmaceuticals are in chronic use among American and European populations. Clinical data on patients using these medications has accumulated for some time. Could there be a medication that is already approved for another use that also has a major benefit for aging? It could be a medication that retards heart disease or prevents cancer. It could even be a medication that has long-term benefits for the cognitive decline that accompanies aging. As an additional benefit of this search strategy, this established medication would already have passed tests for toxicity. Its adoption as an aging treatment could therefore be rapid. It might be an overnight success story, an expeditious launch for the treatment of aging.

Implementing the second strategy would require a considerable research effort, and cooperation from a range of physicians, institutes, and hospitals. Unlike the genomics R&D approach, serendipity is an absolute requirement for success. But the resources required would not be boundless, and the benefits would be considerable.

The basic science of aging, especially the evolutionary biology of aging, suggests that the postponement of human aging would be a difficult but not impossible task for a properly focused research team. The main thing lacking now is some kind of corporate setting in which such a team might work. The problem is thus more one of vision than technology.

15

The Long Tomorrow

As a child I was spooked by Scheherazade's stories from the *Arabian Nights* in which sailors would swim to the bottom of the sea, consort with mermaids for a time, and then surface to find that many years had passed. The sailors that returned from the deep remained young, but everyone around them was old, or dead. Other myths concerning the amelioration of aging involve sex, notably the Taoist tradition. One Taoist recipe for male immortality was having numerous sex partners in just 24 hours, without climaxing. I didn't hear about this as a child.

The most common aging stories involve the ingestion of substances or exposure to them, especially liquid substances. The reversal of aging in the waters of the Fountain of Youth. A vampire's blood that converts you to the immortally undead. The pink liquid from the film *Death Becomes Her*, which made Goldie Hawn and Meryl Streep live forever, albeit with continuing bodily decay. Water from the Holy Grail. The list is endless.

In the depths of human psychology, life is associated with ingestion. This is a legitimate association. We get the foods and the fluids that we need by opening our mouths and then licking, sucking, chewing, and finally swallowing. These are primal experiences. It is natural that fairy tales describe magical liquids with the power to extend life, sometimes indefinitely: the elixir of life.

The concept of an elixir of life fits well with some tendencies that were common in twentieth-century biology. In molecular and cell biology, there is a great deal of interest in finding the specific gene that codes for the particular enzyme that carries out a metabolic task. Many genetic diseases occur because a genetic mutation stymies the production of a single enzyme. It is understandable that some biologists have sought the key genes that regulate aging, the genes that might supply an elixir of life. These researchers often describe their research as the working out of the *regulation of aging* of their organism. With such a viewpoint, it is not difficult to imagine oneself on the trail of the fountain of youth, or at least *the* genetic pathway that controls aging, the molecular elixir of life.

Evolutionary biology gives one an entirely different set of prejudices. Since evolutionists know that aging is an inadvertent effect of multiple evolutionary processes, we understand that aging is not truly regulated. It isn't like the growth and differentiation of the human embryo. It isn't like the transformation from caterpillar to butterfly that occurs in the chrysalis. Those processes are finely regulated. Aging isn't. Indeed, my favorite metaphor for aging is that later life is a landfill. This landfill is replete with the nasty leftovers of early reproduction: clogged arteries, burnt-out livers and kidneys, and abused lungs, among other things. Merely unfortunate genetic effects on later life play a role too. These include biochemical deficiencies that don't matter when you are young, but leave the elderly debilitated. Alzheimer's disease may be caused by an example of this type of genetic junk. Evolutionary biologists know that there will not be a single "magic bullet" or elixir of life to solve all these problems. Aging is instead a many-headed monster. This much is review.

Some time after my failure to emulate Oppenheimer, I returned to the subject of postponing human aging in 1999 with a *Scientific American* article. Over two decades, I have become both more optimistic about the future of human aging and more convinced of the considerable obstacles that lie in the way of progress with human aging.

The most important thing is to understand that there will never be a fountain of youth for human aging. The huge effect of some mutants notwithstanding, especially mutants that reduce fertility or metabolism, there is little prospect of producing a single substance that will postpone human aging on a large scale. Instead of a single magical fluid flowing from a fountain of youth, there will be a cornucopia of youth that will dispense a wide array of products and treatments that together will postpone human aging. Many therapies will be required. It is only when this premise is accepted that progress can be made.

To understand the importance of this starting point, consider the problem of postponing human aging if there *were* a fountain of youth. To have a fountain of youth, the aging process would have to be terminated, and perhaps reversed, by a single substance, an elixir of life. The existence of such an elixir in turn implies a single master pathway in our biochemistry that establishes and coordinates the diverse processes of aging. Let me explain this in detail.

There is no biochemical reason why an elixir that stops aging can't exist. There are many individual processes in our biochemistry that have diverse, important effects: the Krebs cycle of energetic metabolism, DNA replication, the synthesis of proteins, and so on. If these key molecular processes, or pathways, are interrupted at a single point, disaster results. A number of rare disorders are associated with pinpoint disruptions of these pathways, from phenylketonuria, which impairs brain function, to xeroderma pigmentosum, which causes prolific skin cancer.

If aging was another of these disorders, then we could do powerful research on the human genetics and biochemistry of the disorder. Since any particular biochemical pathway in our bodies can be terminated by single mutations, the war against aging would require searching for individuals whose aging pathway has been disrupted. Such people would live much longer than normal, when protected from contagious disease, and be free of the numerous pathologies of aging. They would have little in the way of wrinkles, hypertension, athletic decline, and so on. From studying these people, we could quickly home in on the central aging pathway and then find substances that would stop the aging process, in the same way that supplying insulin alleviates diabetes. The assumption of this scenario that aging is specifically regulated would allow us to postpone it radically with a single biochemical intervention.

Bob Shaw's novel *One Million Tomorrows*, like other works of science fiction and fantasy about aging, is based on this scenario. In the novel's future, scientists discover how to arrest the aging process with finality. This has the ironic side-effect that women are left sexually competent, but men with arrested aging are completely impotent. The only men who remain sexually active are those who allow the aging process to continue. Naturally they are quite, uh, popular with the female population. This trade-off between sex and survival is an interesting echo of actual biological research on the trade-off between reproduction and survival.

Like many of the assumptions of science fiction and popular journalism, this fountain of youth scenario is based on rapid, effective, technological development. If only it were true, we might have a pink liquid to stop aging after a few years of research. Making and selling a great deal of it would then abolish the long-term deterioration of adults. There would be no need for the Social Security Administration, and United States senators could remain in office for centuries instead of decades.

Yet nothing about this scenario is valid. It would be wonderful if the scenarios of mythology and science fiction were true. But they aren't. They can't be. We must instead deal with a messy reality in which progress will be harder, but not impossible.

This fact implies that experiments with humans are unlikely to be good starting points. If aging is not caused by a single pathway, then the complexity of human biochemistry and genetics will make it virtually impossible to sort out human aging from clinical work alone. It *was* possible to figure out Type I diabetes from clinical research with dogs and patients, because the role of insulin in diabetes is fairly simple. That is classic medical research. The postponement of human aging won't fit the classic medical model. It is too complex a problem.

Added to the inherent complexity of aging are two further problems. Humans have long life spans, making numerous tests of interventions that are based on their effect on longevity impractical. In addition, a number of experiments that might be useful, such as lethal stress tests, would be utterly unethical with humans. Any realistic plan for human aging research and development must be based on the use of experimental organisms other than humans. That is, we will need appropriate "models" for the human aging process, animals that can be used as stand-ins for humans.

It was once thought that the ideal model was human cells in vitro. The ultimate cessation of division in these cells, the Hayflick Limit, was offered as the cellular equivalent, or even cellular cause, of human aging. It is now known that the limits to human cell proliferation arise from a well-defined cellular mechanism, telomere shortening. If telomere shortening is stopped, an intervention that we can now perform in vitro, cell proliferation continues indefinitely. Indeed, some cancerous tumors are free of telomere shortening. This makes cell cultures a singularly inappropriate biological system for finding ways to prolong the life span of the entire organism.

For some biologists, any organism that ages is useful for aging research. For example, some scientists study animals with genetic "acceleration" of aging, even progeric children. But these genetic victims might die early for reasons that are unrelated to normal aging processes, rendering them unsuited to research on the control of aging.

Useful model organisms instead must have genetic variants with postponed aging, because those variants must have normal aging mechanisms abrogated, slowed, or minimized.

There are also salient reasons for limiting the list of model species to those that have had their genomes sequenced and their genetics well characterized. If one were to study the aging of tortoises, for example, the accumulation of more genetic information about tortoises would be needed to sort out the aging of the species. One would like to know which genes have variants that make some tortoises live longer than others. It would also be useful for researchers to know the pathology of aging and death in the model organism. But most of these things would not be known in advance in little-studied species. They would all have to be discovered.

Instead of such an arduous research project, the established model organisms offer greater efficiency, greater background knowledge, and established connections to humans. *Drosophila melanogaster* (the lab fruit fly), *Caenorhabditis elegans* (the lab nematode), and *Mus musculus* (the lab mouse) are very powerful model species. All of them have identified genetic variants with postponed aging and sequenced genomes. The advantage of complete genome sequencing is that we can state in reasonably accurate terms which genes are shared between humans, for which we also have a genome sequence, and the model species.

For genes that are shared between humans and a model species, research that discovers a genetic effect on aging in the model species raises the possibility that the corresponding human pathway is involved in aging. This is such an overwhelming research advantage that medically oriented aging research using model species should primarily focus on such common pathways.

But there is more. Well-developed model systems like fruit flies have been dissected in detail both genomically and genetically. We know all their gene sequences and the locations of all their genes. We also know the functional effects of variants of many of the genes of these organisms: their effects on eye color, digestion, locomotion, mating, etc. Researchers have identified and functionally defined variants from more than half the genes in the fly genome. A gene picked at random is likely to be characterized in terms of its functions, not just its DNA sequence. Fortunately, with fruit flies we often have a lot of functional information about the genes that pop out of genomic comparisons, because these genes have often been studied already using other methods. And when we don't already have information about a gene in a fruit fly, we can easily get more.

Even though aging is a process involving many genes, biology developed the tools for handling just such complex biological problems right at the start of the twenty-first century. Knowing nothing else, and we know plenty more than that, the quality of these tools shows that this is a propitious moment for a concerted attack on the problem of postponing human aging. With tools that were not dreamed of 20 years ago, we are given the pleasurable task of deciding on the best strategies for conquering human aging.

Let me begin by criticizing the strategy that I proposed in 1984. Then I argued for a mouse-based research strategy. If we had mice with genetically postponed aging, I was confident that we could proceed from their postponed aging to intervention in the human aging process. I chose mice not because of any familiarity with them; I have never worked with mice. My grounds were those of evolutionary similarity. I knew that mice had to have more genes like those in humans, compared to fruit flies. Therefore, since many genetic loci had to be tried out, the jump from mouse to human would have less chance of failure.

But that isn't important now. The last 20 years have repeatedly seen the scientific exploitation of genetic discoveries from fruit flies, especially their application to human development. The path of inference from these simple model organisms to our complex physiology is not just well beaten. It is a two-lane road paved with a thick asphalt of research publications. If there is anything that a biologist can be confident of, it is the continued interplay between research with fruit flies and research with humans.

To give an example from recent aging research, nematode, fruit fly, rodent, and clinical findings all implicate metabolism in the control of aging. Food turns up in all of these organisms as a major modulator of aging. While it is reasonable to doubt some parts of this story, the overall pattern is clear. The use of food energy affects the rate of aging.

Mice will always have their uses in aging research, even though they are slow experimental systems compared with fruit flies and nematodes. But, contrary to my earlier proposal, they are not essential to making progress on the problem of postponing human aging. Their preeminent role may be more in the late stages of developing interventions, such as testing for drug toxicity or the effects of drugs on mammal longevity, testing that shouldn't be performed clinically at first.

Where does the balance lie? On one hand, we have these wonderful model systems and technologies. On the other hand, we have the multigene complexities of aging, the many-headed monster. What is the best plan of battle?

The first principle must be that the minute dissection of the genetics of single genes is probably inefficient. Traditional biologists will scoff that this is the only way to find out which genes are doing what, but they can't have considered the genomics revolution seriously. The genomics potential of evolutionary experiments is particularly significant.

Natural selection in cultures of fruit flies can easily act on hundreds of genes at the same time. Natural selection acting on outbred animals is not like a traditional geneticist, looking for that single DNA change that will give the right effect. We are outbred animals. Nematodes, yeast, and bacteria aren't, so perhaps geneticists who work with such species can be excused their difficulty with the selection experiments of the fruit fly–aging literature. The aging gene number estimates that Jim Fleming came up with for fruit flies more than a decade ago are like the gene number estimates found by recent genomic research: hundreds of genes are involved in the control of fruit fly aging.

These numbers may cause some of our colleagues to throw up their hands in horror. To me they indicate opportunity. With modern genomic technologies we can test for changes in the expression of thousands of genes simultaneously. We can also look for DNA sequence changes at many genes. Fruit fly populations that have been selected for postponed aging are differentiated for many genes, *any or all of which might be useful for the postponement of human aging*. And we now have the technology for finding most of these genes and characterizing their patterns of expression. Since we have entire genomes sequenced, genes are unlikely to slip between the cracks of our experiments.

The postponement of human aging will not be based on breeding or genetically engineering people to fit some genetic specification. Leaving aside the time that would be needed to do this—probably centuries—it would be both ethically repugnant and technologically stupid. Why build in the last century's genetic conclusions?

The next step should be the transformation of genetic knowledge into pharmacology. Drugs offer the possibility of shifting biochemistry without building any particular genomic technology into our gametes. Drugs can be easily improved and replaced. Drugs too can be approached with a genomics-style strategy, screening them en masse using the smaller model systems, fruit flies and nematodes.

Drugs aren't the only reasonable intervention for the postponement of aging. Some biochemical pathways produce changes that are structurally incorporated in the body. Fruit flies suffer progressive damage with age to the exoskeleton that encases them. This damage results in loss of locomotion, sensation, and fluid. Microsurgery would be the only remedy for most of this damage, if anyone bothered to repair aging fruit flies. Humans are obviously more interested in patching humans, and the potential for surgical repair is now obvious, from orthopedics to heart surgery. Nanotechnology, the next size down from microtechnology, offers the possibility of repairing small blood vessels and neurons, among other tissues. Nanotechnology is not yet deployed, but its potential for assisting aging interventions is vast.

The prospects for discovering how to postpone human aging were negligible for all of human history up until 1980. From 1980 to 2000, those prospects were hopeful, but not outstanding. Since 2000, the year of the sequencing of the human genome, the prospects for postponing human aging have become excellent. We now have all the basic tools that we need, except organizations with the willingness and resources to do the job.

There is a definite paucity of contemporary cultural figures or literary characters who avowedly pursue the treatment of human aging. But there are some. Perhaps the best developed is Lazarus Long, from Robert Heinlein's Future History series. Lazarus appeared first in the aptly named *Methuselah's Children*. In this novel, a privately funded secret group arranges marriages among individuals from long-lived families. Over centuries, they successfully extend human life spans, in much the same way as I extended the life span of my fruit flies by selective breeding. Significantly, when this group goes public with their achievement, in the novel, the reaction of the government is to hunt them down for failing to disclose a simple medical cure for aging that the government imagines is the basis for the great longevities of the group. The longer-lived people escape in a spaceship. These events force the government to initiate research into pharmaceuticals that give ordinary people increased life spans.

There are several assumptions built into Heinlein's story. The first is that governments won't initiate genuine research for the treatment of aging. Private groups have to do it. That seems plausible. His second assumption is that governments and the masses will probably be hostile toward private groups that postpone their own aging. This is a variant of the common "people who want to increase their life spans are evil" plot point, frequent among science fiction novels and movies. I have heard self-appointed "ethicists" speak this way, but I don't know if most people have such opinions. Third, Heinlein assumed that if govern-ments are given enough of a kick in the pants, they too might do some-thing toward postponing human aging. That's probably more of a fantasy than science fiction.

I first read Heinlein's novel in the 1960s, and his main points about the postponement of aging made absolutely no impression on me at the time. Since then I have learned that many people, at least in West-ern countries, are hostile to the treatment of aging.

At the other extreme are a variety of private groups that want the development of new biological technology for the sake of their per-sonal survival. Groups like the Los Angeles Gerontology Research Group (www.grg.org), the American Association for Anti-Aging Medi-cine (A4M), and Alcor, among others, support the development of anti-aging, life-preserving, and related technologies. These groups have little impact on national politics in Western countries, perhaps due to a lack of organization, and perhaps because the prevailing climate in the West is hostile to their goals.

This is obviously a David vs. Goliath situation. When your group of activists is never heard from on op-ed pages or on radio shows, to say nothing of television, you have already lost the debate. Therefore, the more important question about any future Long Tomorrow is how much time will have to elapse before significant technologies that slow aging are allowed to be developed and implemented. Political hostility could make this a very long time indeed.

But the opposite question can also be put, have governments and churches ever succeeded at stifling for good a major technology that would benefit people? Birth control, for example, was a bit of biology that was opposed by many institutions, from governments to churches. Yet it is now widely available in industrialized countries. For this reason, I think that it is still reasonable to hope that eventually the great mass of people will be able to control their aging through pharmaceuticals and medi-cine. Without doubt, many of them will indeed try to do so.

It is no great feat of intellectual courage to predict that we will some day be able to postpone, slow, and otherwise treat human aging. Aging will undergo a revolution. Confused hopes and fears left over from an earlier era will no longer keep us from a new epoch in human health. We will collectively pass through this period of commotion to attain lives of health and opportunity. Tomorrow will be long, and it will be good. But we haven't got there yet.

16

Travels with the Boatman

I didn't grow up with an ambition to save mankind from aging. I was entirely fatalistic about my own aging as a young man, partly because growing old didn't seem so bad in my family. My great-aunt Grace played duplicate bridge into her nineties, an accomplishment that I could barely fathom as a mediocre contract-bridge player. The fact that she suffered from swollen feet seemed like a small burden. I felt very little fear of aging as a youth. Now, of course, I understand better the multifold depredations of biological time that constitute aging. The question is obvious: what does the postponement of death mean to me personally, since it has been the daily work of my laboratory?

My attitudes toward aging and death have been unconsciously marinating over the course of my life. But as of this writing I am just a few months shy of 50 and it is about time that I faced my own prejudices, hopes, and fears about the topic. It won't get easier if I wait for the milestones marked by later decades of life, should I reach them. The readers of this book may well feel that I owe them more candor than I have shown to this point. Here I will try to oblige this interest. However, any scientific colleagues who have read this far might find it the better part of valor to skip this chapter. It will be personal and awkward, in a way private.

I learned about mortality as a child from hair-raising near-death experiences. I have already mentioned my brother Tim almost drowning under ice. Around the same time I slid partway down the side of a sandstone cliff, somehow escaping the final drop to oblivion. My ultimate adventure was caroming off the side of a ski jump into large boulders at the age of 14. I spent months in casts, wheelchairs, and physiotherapy after that. I later developed vertigo and a lack of pleasure in such activities as mountain climbing and downhill skiing. I knew my body could be broken.

I have also already told you about my grandfather Fred Horsey, who lived for an entire century. By contrast, my grandfather Jack Rose died of cancer in 1973, just 74 years old. He had a pronounced fear of doctors dating back to his boyhood in England, and had avoided treatment or diagnosis for chronic abdominal pain. That pain was probably first due to operable primary malignancies, but by the time of exploratory surgery it was already hopeless. There were malignancies everywhere when they opened him up. He had been the picture of health when I had stayed with him in the summer of 1968. By 1973, he was a dried-up husk of his former self. However, I told myself, my grandfather could have lived if he had been better about seeing his doctors. Even with his "early" death, the average age at death of my grandparents was 90. I have since been of the firm opinion that one should always consult doctors, even though they aren't actually demigods able to cure all ills, no matter what some patients, and the worst doctors, think.

Both of my grandmothers died in their early nineties, one quickly and mercifully, the other slowly and painfully. Both retained their faculties to the end. Their sisters and brothers mostly lasted well too.

My parents are still alive and vigorous. They would be the despair of a trust-fund leech. My father is an ace on the tennis court; he and my Aunt Pamela, both in their seventies, have bitter matches in which each tries to prove that they are the better player. My father has a blistering serve. Pam is a competitive horse jumper. A photo of my Aunt Stephanie at 70 featured prominently in a Japanese television program about slowing the aging process.

Starting a few years ago, I developed a wide range of troubling symptoms: numbness, chronic esophageal pain, parasthesia, gross tremor, insomnia, polyuria, and other things too embarrassing to mention here—I had wide-spectrum pathophysiology. I saw a lot of doctors. Brain tumors and multiple sclerosis were among the less attractive diagnoses on offer. Eventually a neurologist convinced me in 2004 that I had peripheral neuropathy. The suspect was one of my long-term prescription drugs. I went off the obvious culprit, and soon started to recover.

I was put through a series of strenuous tests, including a long morning with a neurologist armed with electrical zappers. He checked for damage to both sensory and motor neurons by running current through my body, checking for the responsiveness of motor neurons and the ability to feel pain. Both were intact, unfortunately. Memories of vivisected frogs from my undergraduate physiology labs came back to haunt me.

Halfway through, I said to him, "There are people in San Francisco who would pay good money for this."

The physician's mouth turned tightly upward at the corners, but he said little or nothing.

When he seemed to have finished, I asked him if it was all over.

"We can stop now," he said. "Unless you would like me to continue?"

I declined his kind offer.

But there would be worse.

During radiological work on the extent of the damage to my spinal cord, an MRI was taken. It seemed to reveal a vertebral tumor. Vertebral tumors are often caused by secondary malignancies of metastasizing cancers from prostate glands and other tissues. When you have such metastasis, there is no curative treatment. Palliation is the best that physicians offer. Death may come within months or at best a few years from diagnosis. So far, though, there is no metastasis.

Two further events confronted me with mortality. Janice Nassany, a long-time friend who had diabetes, died of heart failure. She had been in ill health for decades, losing fingers, toes, and most of a foot. She spent a lot of time in wheelchairs. Always one of the most upbeat people you could ever hope to meet, and a big music fan like myself, she was about the last person whose burial I wanted to attend. But there I was, in 2004, seeing her coffin beside a grave dug out of a Newport Beach hillside. She was about 50, like me.

Then Larry Mueller, my stalwart colleague and just a few years older, was diagnosed with kidney cancer in October 2004. One of his kidneys had a large tumor. It shocked everyone because Larry is a former track-and-field athlete who is still in excellent condition. Surgeons removed the entire kidney right away. There were some very tense days until the histology of the excised kidney suggested an excellent prognosis. Larry will be counting fruit flies for decades to come.

Death was no longer a nebulous possibility to be faced decades in the future. Death had become something that threatened my peers, something that might happen to myself, and soon. Now I had to answer the question, what did death really mean to me?

My answer is that death is not the most fearsome thing that I have faced in my life. Indeed, I have endured debacles that I feel are less appealing than the prospect of my own dissolution. But those misfortunes were unforeseen, and—like the Ancient Mariner—I was never offered the choice of death versus death-in-life. The Fates tossed their dice for me, and I have endured the outcome of their gaming.

Yet to finally die does mean the end of life. So what does the continuation of life fundamentally mean to me? How much does one really lose if one dies at 50 or 60 years of age?

I can only answer this question in terms of my own values and experiences. I have had a substantive academic career, and I don't feel that I have wasted my life having a scientific vocation. I certainly can't regret having spurned medicine or the law. My accomplishments mean a lot to me, even though I know that science is a fluid thing, like an endless soccer game. I am content with the feeling that I usually kicked the ball farther up the field during my playing career. Occasionally I have scored goals.

The greatest regret is of course the personal. The most human side of my life has been dominated by loss and catastrophe, the details of which I will not recount further than I have already. I think few people face oncoming death without having regrets about their relationships, the words they wish had never parted their lips. I will never understand the persecution of those who try to perpetuate their lives, people who may only want more of an opportunity to get things right. The moralists who would deny others the chance of personal renewal or reconciliation from a longer life could be in the grip of thanatos, Freud's death instinct. Death must be the comforting blanket for their sickness.

The river Styx defines the upper reaches of Hades. It is variously rendered as slow and peaceful or as swift and unfordable. The only way across is to use Charon's boat. Sometimes Charon is a wispy almost benign figure. Others render him more concretely as a skeletal figure with a terrifying visage. His dress is always classical, usually robes. Sometimes he holds a scythe. Charon is an immortal, but not quite a god. He requires payment for the trip, and this is the mythological justification for the coins placed in the mouths of the dead. "To pay the boatman . . ."

I have worked on the banks of the river Styx for 30 years, thanks to my research. They have been years of great fulfillment for me as a scientist, though profoundly difficult for me as a creature of human feelings and desires. I will be crossing the river sooner or later, and I hope the boatman is good to me during the voyage. I think I have a coin that will be of use in his travels.

Acknowledgments and Disclosures

Though I made an abortive attempt to write a book like this with Randy Black in the early 1990s, the chief instigator of this particular effort was Lisa Adams, who worked with me as my agent in the first part of the year 2000. Later that year, Faith Hamlin took over as my agent, and has since guided the book toward its final publication with Oxford University Press. Faith has been a great help and inspiration over the last four years, and I am very grateful to her.

Peter Prescott has served as my editor at OUP. I am beholden to him for providing a home for *The Long Tomorrow*.

Of course there wouldn't have been anything to write about if it hadn't been for Brian Charlesworth and his determination to get me to work on aging when I thought it was a fool's errand. Brian handed me my professional career, and I feel that I have never properly expressed my gratitude to him. I view this book as one long thank-you letter to him.

Not to be overlooked are the students, technicians, postdoctoral fellows, colleagues, and advisors who also helped me with the research I describe here. They number more than 400, and I couldn't reasonably list them all here. A fair number of them are listed as the authors and the acknowledged assistants and colleagues for the papers collected in *Methuselah Flies* (2004, World Scientific Press, Singapore). However, here I would like to acknowledge my debt to the late John Maynard Smith, Deborah Charlesworth, the late William Hamilton, Amanda Simcox, James F. Crow, Bill Atchley, Lefteri Zouras, Phil Service, Ted Hutchinson, Rich Lenski, Joe Graves, Leo Luckinbill, Robert Arking, Tom Johnson, Armand Leroi, Adam Chippindale, Linh Vu, Ted Nusbaum, Margaret Archer, Hardip Passananti, Denise Deckert-Cruz, the late Robert Tyler, Francisco Ayala, Jim Fleming, Greg Spicer, Roger Garrison, Tim Bradley, Minou Djawdan, Caleb Finch, Allen Gibbs, Eric Toolson, Larry Mueller, Larry Harshman, Margarida Matos, John

Tower, Dan Borash, Jay Phelan, Henrique Teotónio, the late Roy Walford, Mark Drapeau, Adrienne Williams, Donna Folk, Tony Long, and Casandra Rauser, in roughly chronological order. I apologize to those left off this list, particularly my many dedicated undergraduate research students.

My work on aging has been substantially supported by a British Commonwealth Scholarship, a NATO Science Fellowship, a Canadian government University Research Fellowship, Dalhousie University, grants from the Natural Sciences and Engineering Research Council of Canada, grants from the National Institute on Aging, grants from the National Science Foundation, the family and estate of Robert H. Tyler, as well as the University of California–Irvine.

I hereby disclose that I was a paid consultant to MRx Biosciences, Incorporated, for several years in the 1990s. Since then I have actively worked on multiple corporate start-ups in the aging field, though without material compensation. I am at present the chief scientific consultant to Longevity Biopharmaceuticals, Incorporated, though I am still not paid for my consulting with them. I hope to be paid for my work with them in the future and I have agreed to receive a small percentage of the stock of the company, which remains an unfunded start-up as of this writing. I would also like to acknowledge some of my past and present corporate colleagues, particularly Robin Bush, Ted Nusbaum, Ralph Andrews, Bill Andrews, Parviz Sabour, Owen McGettrick, Greg Stock, Peter Donald, Linda Howell, Tony Long, Larry Mueller, Cristina Rizza, Marc Rofeh, and Mahtab Jafari. This list is again quite incomplete, and I apologize to those I have omitted.

This book was read in draft by a great number of patient friends and colleagues. I would especially like to thank Faith Hamlin, Lisa Adams, Jennifer Ehman, Greg Benford, Jay Phelan, Dana Ohler, Parvin Shojaeian, Margarida Matos, Susan Rattigan, Gordon Lithgow, Natasha Vita-More, Casandra Rauser, Mark Drapeau, Theresa Hansen, David Kekich, Stephen Coles, Larry Mueller, Cristina Rizza, Blanca Cervantes, and, of course, Peter Prescott for their suggested revisions and encouragement. Again, I apologize to the readers I have overlooked. It should be said that I did not implement all the advice I received, and the final form of the text is my responsibility.

Glossary

adaptation A biological attribute that increases net reproduction (or fitness); or, a product of natural selection that increases net reproduction; or, the process of natural selection by which fitness is increased.

age-class A group of individuals that share a range of ages; e.g., humans between the ages of 10 and 20 years of age.

age-structure A population's make-up in terms of its age-classes.

aging A protracted decline with age in survival or fecundity even under conditions of excellent care.

allele A particular variant form of a gene; e.g., the allele for brown eyes in humans.

asexual reproduction Reproduction without genetic exchange with another organism; may involve recombination among the genes of one genome; may involve seeds, buds, branching, or splitting into two symmetrical parts.

biological species Groups of populations that do not interbreed with other populations.

castration The removal of the reproductive organs of an organism.

chromosome A long string of DNA that encodes an organism's genes.

clone A group of genetically identical organisms, usually produced from a single parent.

cohort A group of individuals that begin life at the same time.

comparative method The inference of adaptation from the comparison of species.

cost of reproduction A reduction in survival resulting from reproduction.

cystic fibrosis A genetic disease that produces abundant mucous leading to persistent lung infections and other health problems.

Darwinism   The scientific theory of biological evolution directed by natural selection.

dauer   A stage that postpones reproduction but has better survival under stress.

diploid   When all chromosomes are present in exactly two copies.

drones   Adult males in Hymenopteran eusocial insect societies.

electrophoresis   The separation of proteins using charged gels.

eukaryote   An organism with a nucleus distinct from the cytoplasm and other characteristic organelles, such as mitochondria or chloroplasts.

evolution   A change in the genetic composition of a population over one or more generations.

external fertilization   When syngamy arises from the fusion of gametes released into the surrounding medium, instead of occurring inside the body of the female.

fecundity   A female's quantitative production of eggs.

fertility   The number of viable offspring produced by their parent(s).

fission   The division of a cell or a multicellular organism to form two cells or organisms, respectively.

fitness   The net reproduction of an organism over a complete life cycle.

force of natural selection   The intensity of natural selection acting on an age-specific change in survival probability or fecundity.

gamete   A cell that forms zygotes by syngamy.

gene expression   The transcription of a gene's DNA sequence.

generation time   The average length of time between the parents of each generation.

genetic disease   A disorder caused by a single genetic change.

genetic engineering   The deliberate modification of the genome of an organism's cells, whether somatic or gametic, using molecular technology.

genome   The complete set of chromosomes; all the DNA in a cell.

genotype   The specific alleles at a locus or group of loci.

geriatrics   The medical specialty concerned with the treatment of the elderly.

germline engineering Genetic engineering of the cells that make gametes.

gerontology The scientific study of aging.

gonad An organ that makes gametes.

group selection Natural selection between entire groups.

haploid A full set of chromosomes, without duplication.

herbivores Animals that mostly eat plant matter.

hermaphrodite An organism that produces both sperm and eggs, or their equivalent.

heterozygosity When a diploid locus has two different alleles; or, the fraction of loci that are heterozygous.

homozygosity When a diploid locus has two copies of the same allele.

hunter-gatherer The human way of life before the advent of agriculture.

Huntington's disease A dominant genetic disease characterized by progressive loss of central nervous system function.

hybrid vigor Increased fitness in offspring produced from crosses of differentiated stocks, breeds, varieties, or selected lines.

identical twins Individuals that share a mother and all their gene sequences.

immortality, biological Stable survival rates irrespective of age; may not apply to all life-cycle stages.

inbred line Organisms that have been subjected to high levels of inbreeding for many generations.

inbreeding Mating of close biological relatives.

internal fertilization When fertilization occurs inside an animal's body.

iteroparous Life histories that have repeated bouts of reproduction.

just-so stories Plausible biological theories that lack important details.

Lamarckism In part, the view that the effects of use or disuse are transmitted to offspring.

life cycle The pattern of birth, development, reproduction, and death of a species.

life history   The quantitative pattern of survival and reproduction of an organism, genotype, or species.

maximum longevity   The greatest known longevity of a species.

meiosis   The process by which gametes are produced by eukaryotes.

natural selection   The differential net reproduction of particular genotypes.

net reproduction   The average total reproductive output of a genotype, estimated for all newborns, whether they live or die.

outbred   When breeding rarely involves close relatives.

parental care   When parents aid their offspring.

parthenogenesis   The production of offspring without genetic fertilization, though possibly with copulation.

phenotype   A characteristic of an organism.

phenotypic selection   Selection that depends on a phenotype.

population   A group of organisms that interbreed frequently.

queen   In social insects, the female who produces all the eggs in a familial colony.

recessive   When an allele does not affect the phenotype unless it is present in two copies.

recombination   The reassortment of alleles among gametes due to the shuffling of chromosomes or their parts.

semelparous   When the life history has just one bout of reproduction.

senescence   Usually a synonym for aging.

sexual reproduction   Reproduction in which gametes form zygotes by syngamy.

sexual selection   When selection acts on mating success.

social insect   An insect species in which kin selection favors the production of less fertile workers who aid the parent(s) of a colonial group.

soldier   A sterile social insect "caste" that attacks prey and defends a social insect colony.

somatic engineering   The genetic engineering of cells that don't make gametes.

sperm Mobile gametes that fertilize egg cells.

syngamy The fusion of two gametes to form a zygote.

Taoism An ancient Chinese cosmological system with religious and protoscientific elements.

Tay-Sachs disease An incurable recessive genetic disease caused by a deficiency in fat metabolism resulting in the progressive poisoning of the brain; death occurs in childhood.

testes Male gonads.

transcription The production of RNA from genomic DNA.

transformation The insertion of exotic DNA molecules into a genome.

translation When RNA directs the assembly of a protein.

variation, genetic When there is more than one allele per locus in a population.

variance, phenotypic A measure of the variability between organisms for a quantitative character.

vegetative reproduction Reproduction without any form of sex.

viability Probability of survival from zygote to adulthood.

worker A sterile social insect "caste" that cares for other members of the colony, especially the queen.

X chromosome A sex chromosome present in two copies in mammalian females and one copy in males.

Y chromosome A sex chromosome that is normally found only in mammalian males.

yeast A fungal species that is used to bake bread and culture wine.

zygote The cell that starts development, usually a diploid cell produced by syngamy of haploid egg and sperm.

Bibliographic Essay

Formal citation has to be one of the most sterile and intimidating features of the academician's art. A particular irritant is the citation that merely gives a source, without indicating how specifically pertinent the citation is, or indeed how useful it might be for the curious reader. In this bibliographic essay, I continue a practice I started with *Darwin's Spectre* (1998; Princeton University Press): presenting my citations with an explanation of my intentions in citing them, as well as some indication of how useful it would be for readers to look up the reference material for themselves.

I have divided my bibliographic essay into two parts. The first part gives generally useful references that underscore, extend, or illuminate the material of this book. The second part discusses citations that are specific to the individual chapters.

General Reference Material

As this is a work that attempts to popularize science, it has as its foundation the actual science that it discusses only at second hand. Those foundations are well worth the time of the reader who is seriously interested in the topic. Where aging is concerned, there is one central source of documented information: *Longevity, Senescence, and the Genome* by Caleb E. Finch, first published in 1990 by the University of Chicago Press. I have some disagreements with a few passages in this work, but everyone in the field knows that it is the single best source of "facts" concerning aging.

The journals *Experimental Gerontology, Journal of Gerontology, Mechanisms of Ageing and Development, Aging Research Reviews, Rejuvenation Research,* and *Biogerontology* regularly update the experimental substratum of gerontology, the scientific study of aging. There is also a spectrum of other periodicals that subsist on the periphery of the field of aging research.

A book of this kind is naturally dominated by the opinions and prejudices of its author. It is also somewhat deformed by the limitations imposed by the goal of communicating to a wide audience. Fortunately, there are also books that present this field without such compromises. For a definitive statement of the mathematical theory that underlies evolutionary research on aging, see *Evolution of Age-Structured Populations* by Brian Charlesworth (1st edition, 1980; 2nd edition, 1994; both published by Cambridge University Press). For lesser mortals, my *Evolutionary Biology of Aging* (M. R. Rose; 1991 & 1994 editions; Oxford University Press) is a combination of the elementary theory for the evolution of aging with relevant experimental research. A great deal has happened since the publication of this last book, but its basic ideas remain relevant. A still more elementary work of this kind is Steven N. Austad's *Why We Age, What Science Is Discovering about the Body's Journey Through Life* (1999; Wiley). Austad is a "Great Communicator" of aging research.

As to the details of research with fruit flies, the book *Methuselah Flies, A Case Study in the Evolution of Aging* (M. R. Rose, H. B. Passananti, M. Matos, eds.; 2004; World Scientific Publishing) covers the research of my lab in detail, as well as referring in passing to other fruit fly research on aging. *Genetics and Evolution of Aging* (M. R. Rose & C. E. Finch, eds.; 1994; Kluwer) covers a wide spectrum of research on aging. Research on aging in yeast and nematodes is introduced nicely by Lenny Guarente in his *Ageless Quest: One Scientist's Search for Genes That Prolong Youth* (2002; Cold Spring Harbor Laboratory Press). Research on "cell aging" is presented exhaustively by Michael Fossel in *Cells, Aging, and Human Disease* (2004; Oxford University Press). Michael Fossel was also the founding editor of the *Journal of Anti-Aging Medicine*, a heroic effort to jump-start research on postponing or slowing human aging. This journal is now known as *Rejuvenation Research*, and it is edited by the charismatic Aubrey de Grey, whose dedication to the "cause" is almost as formidable as his appearance.

There is a wide array of journalistic, declamatory, and variably reliable books on aging, from *Stop Aging Now! The Ultimate Plan for Staying Young and Reversing the Aging Process* (Jean Carper; 1995; HarperCollins) to *The Quest for Immortality, Science at the Frontiers of Aging* (S. Jay Olshansky & Bruce A. Carnes; 2001; Norton). These books range from inappropriately optimistic to gratuitously negative.

BIBLIOGRAPHIC ESSAYS FOR INDIVIDUAL CHAPTERS

Chapter 1: The Sphinx and the Rabbi

This Philadelphia meeting gave rise to a book, *The Fountain of Youth; Cultural, Scientific, and Ethical Perspectives on a Biomedical Goal* (S. G. Post & R. H. Binstock, eds.; 2004; Oxford University Press). My essay in this book is called "The Metabiology of Life Extension." Arking's essay is "Extending Human Longevity: A Biological Probability." The theologian Diogenes Allen contributed an epilogue: "Extended Life, Eternal Life: A Christian Perspective," while Rabbi Neil Gillman states his views in "A Jewish Theology of Death and the Afterlife." The entire event was filmed and has been available at the Web site of the Templeton Foundation. The quotations given in chapter 1 are taken directly from the filmed utterances of the participants. However, for a full presentation of the views of the participants, it is advisable to consult their essays in the meeting book.

Chapter 2: Maynard Smith's Shirts

The proceedings of the Sherbrooke meeting were published as *Mathematics and the Life Sciences. Proceedings 1975* (D. E. Matthews, ed.; 1977; Springer-Verlag). John Maynard Smith (1920–2004) published many books in his life, including *Mathematical Ideas in Biology* (1968; Cambridge University Press), *Models in Ecology* (1974; Cambridge University Press), and *The Evolution of Sex* (1978; Cambridge University Press). He never published a book on aging, even though it was his research on aging that first made him prominent among English biologists. He did, however, contribute a summative paper on the evolution of aging to *Evolution in Health & Disease* (Stephen C. Stearns, ed.; 1999; Oxford University Press): "The evolution of non-infectious and degenerative disease," together with a crew of co-authors.

An example of my work with ecological simulation is "Using sensitivity analysis to simplify ecosystem models" (M. R. Rose & R. Harmsen; 1978; Simulation 31: 15–26).

For examples of both the dazzling penetration of Richard C. Lewontin and his predilection for Marxist holism, see his book *The Genetic Basis of Evolutionary Change* (1974; Columbia University Press). The first five chapters are an excellent history of evolutionary biology since 1930, but in chapter 6 he pushes for a holistic theory of population genetics that has since been clearly falsified. Lewontin and Richard Levins present their ideological version of biology in *The Dialectical Biologist* (1985; Harvard University Press).

Edward O. Wilson has been one of the more prolific biologists of our time. Among his books are *The Insect Societies* (1971; Harvard University Press), *Sociobiology: The New Synthesis* (1975; Harvard University Press), and *On Human Nature* (1978; Harvard University Press). His view of his relationship with Lewontin is given in *Naturalist* (1994; Island Press).

Chapter 3: Cell Gang

Elie Metchnikoff published his ideas in *The Nature of Man* (1904; Heinemann) and *The Prolongation of Life: Optimistic Studies* (1908; Putnam). Paul Ewald's best book so far is his *Evolution of Infectious Disease* (1994; Oxford University Press). Caleb E. Finch and Eileen M. Crimmins have just published a good article on the topic of inflammation and aging, though their evidence is indirect ("Inflammatory exposure and historical changes in human life-spans," 2004, *Science* 305: 1736–739).

G. P. Bidder published the growth-limitation theory in "The mortality of plaice" (1925, *Nature* 115: 495–96) and "Senescence" (1932, *British Medical Journal* 115: 5831). Alex Comfort then destroyed it in "The longevity and mortality of a fish (*Lebistes reticulatus* Peters) in captivity" (1961; *Gerontologia* 5: 209–22) among other publications (reviewed in his *The Biology of Senescence, Third Edition*, pp. 88–105; 1979; Elsevier).

One of the best sources for the cell and molecular fashions in aging research is *Modern Biological Theories of Aging* (Huber R. Warner, Robert N. Butler, Richard L. Sprott, & Edward L. Schneider, eds.; 1987; Raven Press). This book came out just as the dominance of cell biology in aging research was about to collapse. The original publications of the somatic mutation and error catastrophe theories were L. Szilard (1959) "On the nature of the aging process" *Proceedings of the National Academy of Science, USA* 45: 30–45, and L. E. Orgel (1963) "The maintenance of the accuracy of protein synthesis and its relevance to ageing" *Proc. Natl. Acad. Sci,. USA* 49: 517–21, respectively.

The failure of Maynard Smith's lab to induce an error catastrophe was published as F. Dingley and J. Maynard Smith (1969) "Absence of a life-shortening effect of amino-acid analogues on adult *Drosophila*" *Experimental Gerontology* 4: 145–49.

The classic Hayflick Limit publications are L. Hayflick and P. S. Moorhead (1961) "The serial cultivation of human diploid cell strains" *Experimental Cell Research* 25: 585–621, and L. Hayflick (1965) "The limited *in vitro* lifetime of human diploid cell strains" *Experimental Cell Research* 37: 614–36.

One historical account of Alexis Carrel's work is given by Mark Benecke in his book *The Dream of Eternal Life; Biomedicine, Aging, and Immortality* (2002, pp. 11–13, Columbia University Press). Hayflick's own interpretation of Carrel's work is supplied in his book *How and Why We Age* (1994, pp. 112–15, Ballantine). The impact of Carrel's "finding" on aging research generally is dramatically illustrated by Raymond Pearl in his *Biology of Death* (1922, p. 62, Lippincott), who takes the finding as a "demonstration of the potential immortality of somatic cells, when removed from the body for a far longer time than the normal duration of life." Thus, before Hayflick, aging was not thought of as a problem of cell biology.

I made the Hayflick Limit eponymous in my 1991 *Evolutionary Biology of Aging*, pp. 127–31. Professor Hayflick has never complained about my presumption.

Chapter 4: The Force

J. B. S. Haldane is not only a figure in a variety of books about twentieth-century English culture and politics, he has also appeared in several novels, *Antic Hay* being mentioned in the text. Another is the novel where he appears as himself by name, *The Cambridge Quintet: A Work of Scientific Speculation* (John L. Casti, 1998, Perseus). The book in which he presents the biological ideas that Aldous Huxley used is *Daedalus, or Science and the Future* (1924, Dutton). The book containing Haldane's discussion of aging and Huntington's chorea is *New Paths in Genetics* (1941, Allen & Unwin).

The two Medawar publications are "Old age and natural death" (1946; *Modern Quarterly* 1: 30–56) and *An Unsolved Problem of Biology* (1952; H. K. Lewis). The second piece has been reprinted in other works, as well. I should warn the reader that several of Medawar's seemingly plausible arguments are actually incorrect. The evolutionary theory of aging wasn't really well defined until Charlesworth's 1980 book, mentioned at the outset of this bibliography.

In the 1950s, George C. Williams was thinking along the same lines as Medawar, but it wasn't until he submitted his 1957 paper ("Pleiotropy, natural selection, and the evolution of senescence" *Evolution* 11: 398–411) for publication that he learned about Medawar's earlier publications. Like Medawar, Williams offered some dubious formal arguments, but his intuitive take on the evolution of aging was probably the best the field had before 1980.

To complete the trio of early theoretical efforts, we have W. D. Hamilton's "The moulding of senescence by natural selection" (1966; *Journal of Theoretical Biology* 12: 12–45). This paper, like other Hamilton publications, is a work of pure genius. However, it is not for the faint of heart. In addition to a prose style that can only be called magisterial Oxbridge, this paper contains remarkably poor mathematical notation. Amazingly, when Hamilton reprinted this article in his *Narrow Roads of Geneland, Volume 1, Evolution of Social Behaviour* (1996, Freeman), he did *not* clean up the mathematical notation. Therefore, I do not recommend that nonprofessionals try to use this article.

Chapter 5: Goon Show Einstein

I have already discussed Maynard Smith's books in this essay. A fair example of Paul Harvey's contribution to biology is provided by *The Comparative Method in Evolutionary Biology* (P. H. Harvey & M. Pagel; 1991; Oxford University Press). This book has influenced an entire generation of evolutionary biologists.

Albert Einstein was one of the few great scientists who, like Darwin, could write well for the general public: for example, his book *The Meaning of Relativity* (1950; 3rd edition; translated; Princeton University Press). I haven't looked at it since I was 12, but the feeling of Einstein's powerful mind still haunts me.

Chapter 6: Tiny Methuselahs

Joseph Needham's *Science and Civilization in China* (publication of its many volumes began in 1954, with a growing number of contributors;

Cambridge University Press) is one of the greatest feats of publication in modern times. It is no less than a complete distillation of Chinese material knowledge, with elements of Chinese philosophy, written for a Western audience.

The first J. M. Wattiaux paper I read was "Cumulative parental effects in *Drosophila subobsucra*" (1968; *Evolution* 22: 406–21). Later I read "Parental age effects in *Drosophila pseudoobscura*" (1968; *Experimental Gerontology* 3: 55–61).

The edition of Francis Bacon's aging book that I have read is *Historia Vitae et Mortis* (translated English edition; 1889; Longman).

The effect of delayed reproduction on the future of human aging is discussed in "Evolution of human lifespan: past, future, and present" (M. R. Rose & L. D. Mueller; 1998; *American Journal of Human Biology* 10: 409–20).

Chapter 7: The Postman Rings Again

The publication data for my maiden paper on aging with Brian was *Nature* 287: 141–42.

Several generations of scientists have learned their quantitative genetics from Falconer's *Introduction to Quantitative Genetics* (1st edition: 1960; Oliver & Boyd). It is the most lucid advanced science textbook I have ever read.

John Maynard Smith's 1958 paper ("The effects of temperature and of egg-laying on the longevity of *Drosophila subobscura*"; *Journal of Experimental Biology* 44: B20–22) is a neglected scientific classic. Biology undergraduate students should be required to read it. This paper is discussed in most detail in chapter 8.

James F. Crow's greatest work is *An Introduction to Population Genetics Theory* by J. F. Crow & M. Kimura (1970; Harper & Row), though the term *Introduction* is a misnomer. It was the most advanced textbook in the field when it was published.

The two 1984 papers that really got the field of fruit fly research on aging up and running were my "Laboratory evolution of postponed senescence in *Drosophila melanogaster*" (*Evolution* 38: 1004–10) and "Selection for delayed senescence in *Drosophila melanogaster*" (L. S. Luckinbill, R. Arking, M. J. Clare, W. C. Cirocco, & S. A. Buck; *Evolution* 38: 996–1003). These papers appeared side by side, which was felicitous, although the Luckinbill paper preceded mine in the journal despite its later dates of submission and acceptance, contrary to academic convention, which has caused oddities of citation in the literature on aging.

Chapter 8: Cheshire Cat Cost

A modern-day example of the amelioration of fruit fly aging in conjunction with reduced female fecundity is "Long-lived *Drosophila* with over-expressed dFOXO in adult fat body" (M. E. Giannakou, M. Goss, M. A. Jünger, E. Hafen, S. J. Leevers, & L. Partridge; 2004; *Science* 305: 361).

An example of Phil Service's work on male fertility and the postponement of aging is P. M. Service & A. J. Fales, "Evoluton of delayed reproductive senescence in male fruit flies: sperm competition" (1993; *Genetica* 91: 111–25).

Needham's *Science and Civilization in China* is a good reference for Taoist practices.

Finch's 1990 book, described at the outset of this bibliographic essay, is an excellent source of information concerning castration and aging.

In addition to Finch's tome, Comfort's 1979 book on aging has a great deal of data concerning species life spans. It also provides an excellent review of the empirical issues involved in collecting good data on aging in diverse species.

The Chippindale papers that cover the issue of the cost of reproduction are as follows: A. K. Chippindale, A. M. Leroi, S. B. Kim, & M. R. Rose, "Phenotypic plasticity and selection in *Drosophila* life-history evolution. I. Nutrition and the cost of reproduction" (1993; *Journal of Evolutionary Biology* 6:171-193) and A. K. Chippindale, A. M. Leroi, H. Saing, D. J. Borash, & M. R. Rose, "Phenotypic plasticity and selection in *Drosophila* life-history evolution. 2. Diet, mates and the cost of reproduction" (1997; *Journal of Evolutionary Biology* 10: 269–93).

For example, James T. Giesel did not infer the existence of genetic trade-offs because of inbreeding. See for example "Genetic co-variation of survivorship and other fitness indices in *Drosophila melanogaster*" (1979; *Experimental Gerontology* 14: 323–28). I discuss this type of work, and experimentally challenge it, in "Genetic covariation in *Drosophila* life history: Untangling the data" (1984; *American Naturalist* 123: 565–69).

The search for the key to the Cheshire Cat effect on fly reproduction is summarized in the following 1994 publications: A. M. Leroi, A. K. Chippindale, & M. R. Rose, "Long-term laboratory evolution of a genetic trade-off in *Drosophila melanogaster*. I. The role of genotype x environment interaction" (*Evolution* 48: 1244–257) and A. M. Leroi, W. R. Chen, & M. R. Rose, "Long-term laboratory evolution of a genetic trade-off in *Drosophila melanogaster*. II. Stability of genetic correlations" (*Evolution* 48: 1258–268). We then generalized the idea in "Laboratory evolution: The experimental wonderland and the Cheshire Cat Syndrome" (M. R. Rose, T. J. Nusbaum, & A. K. Chippindale; 1996; pp. 221–41 in *Adaptation*, M. R. Rose & G. V. Lauder, eds., Academic Press).

The rediscovery of the Cheshire Cat effect in aging was published as "Metabolism, flight performance, and fecundity of *Indy* long-lived mutant flies" (J. H. Marden, B. Rogina, K. L. Montooth, & S. L. Helfand; 2003; *Proceedings of the National Academy of Science, USA* 100: 3369–373). Though a former UCI student is one of the co-authors of this paper (KLM), and both studies concern aging and used the same experimental organism, this rediscovery is apparently independent, a striking illustration of the one-way mirror that impairs communication between the two cultures of biology, molecular and Darwinian.

Chapter 9: Birds and Bees

Aristotle's pioneer work on aging is *On the Length and Brevity of Life*. The original publication data are hard to obtain.

As before, the best information on comparative patterns of aging comes from the Finch and Comfort books.

For a first taste of the housefly flight and aging literature, see "Effect of experimentally prolonged life span on flight performance of houseflies" (R. S. Sohal & J. H. Runnels; 1986; *Experimental Gerontology* 21: 509–14).

Finch (e.g., 1990) is particularly interested in long-lived species of fish and other groups.

The idea of comparing the two sexes is discussed further in my *Evolutionary Biology of Aging*.

The consummate study of red deer field biology is *Red Deer: Behavior and Ecology of Two Sexes* (T. H. Clutton-Brock, F. E. Guiness, & S. D. Albon; 1982; University of Chicago Press).

Chapter 10: Deadly Serendipity

Phil's groundbreaking physiology paper was "Resistance to environmental stress in *Drosophila melanogaster* selected for postponed senescence" (P. M. Service, E. W. Hutchinson, M. D. MacKinley, & M. R. Rose; 1985; *Physiological Zoology* 58: 380–89). He followed this paper with "Physiological mechanisms of increased stress resistance in *Drosophila melanogaster* selected for postponed senescence" (1987; *Physiological Zoology* 60: 321–26).

The stress-selection experiments were reported in "Selection for stress resistance increases longevity in *Drosophila melanogaster*" (M. R. Rose, L. N. Vu, S. U. Park, & J. L. Graves; 1992: *Experimental Gerontology* 27: 241–50).

The big analysis of the role of stored calories in starvation resistance was published as "Metabolic aspects of the trade-off between fecundity and longevity in *Drosophila melanogaster*" (M. Djawdan, T. Sugiyama, L. Schlaeger, T. J. Bradley, & M. R. Rose; 1996; *Physiological Zoology* 69: 1175–195).

Among the publications concerning the physiology of fly desiccation are "Physiological mechanisms of evolved desiccation resistance in *Drosophila melanogaster*" (A. G. Gibbs, A. K. Chippindale, & M. R. Rose; 1997; *Journal of Experimental Biology* 200: 1821–832) and "Postponed aging and desiccation resistance in *Drosophila melanogaster*" (D. Nghiem, A. G. Gibbs, M. R. Rose, & T. J. Bradley; 2000; *Experimental Gerontology* 35: 957–69). For more articles on the topic and an overview of this work, see part II of *Methuselah Flies*.

Chapter 11: One Can't Be Too Rich or Too Thin

This literature is surveyed by Comfort (1979), Finch (1990), and my *Evolutionary Biology of Aging*. Our best experimental work in this area is summarized in the two Chippindale papers on phenotypic plasticity. Again, *Methuselah Flies* gives a useful overview, in its part III.

For more on the immediate effects of diet on fruit fly survival, see "Dietary restriction in *Drosophila*" (C. L. Rauser, L. D. Mueller, & M. R. Rose; 2004; *Science* 303: 1610–611).

One paper on the metabolic rate question that does not appear in *Methuselah Flies* is "Does selection for stress resistance lower metabolic rate?" (M. Djawdan, M. R. Rose, & T. J. Bradley; 1997; *Ecology* 78: 828–37).

For an introduction to the physiology of dietary restriction in rodents, see "Food restriction in rodents: An evaluation of its role in the study of aging" (E. J. Masoro; 1988; *Journal of Gerontology* 43: B59–64).

For a variety of views on Roy Walford's Biosphere II adventures, including a wide range of academic citations, see the special number of *Experimental Gerontology* (vol. 39, no. 6) published in his honor shortly after his death. This special number also includes several personal descriptions of the man and his impact on the scientists who knew him.

See *The Okinawa Program* (B. J. Willcox, D. C. Willcox, & M. Suzuki; 2001; Three Rivers Press) for an example of the Willcox Okinawa publications.

One of the best books on the actual numbers pertaining to human aging is *Human Longevity* (D. W. E. Smith; 1993; Oxford University Press).

For more on the search for a pill that mimics the effects of dietary restriction, see "The serious search for an anti-aging pill" (M. A. Lane, D. K. Ingram, & G. S. Roth; 2002; *Scientific American* 287(2): 36–41).

Chapter 12: Many-Headed Monster

A summary of the quantitative gene-number research is provided in "Quantitative genetic analysis of postponed aging in *Drosophila melanogaster*" (E. W. Hutchinson & M. R. Rose; 1990; pp. 65–85 in *Genetic Effects on Aging II*, D. E. Harrison, ed.; Telford Press).

A summary statement of the Maynard Smith lab's work on protein synthesis in fruit flies is "Protein turnover in adult *Drosophila*" (J. Maynard Smith, A. N. Bozcuk, & S. Tebbutt; 1970; *Journal of Insect Physiology* 16: 601–03).

The paper we eventually published from the work at the Linus Pauling Institute was "Two-dimensional protein electrophoretic analysis of postponed aging in *Drosophila*" (J. E. Fleming, G. S. Spicer, R. C. Garrison, & M. R. Rose; 1993; *Genetica* 91: 183–98).

The gene-chip paper on fly aging was "Genome-wide transcript profiles in ageing and calorically restricted *Drosophila melanogaster*" (S. B. Pletcher, S. J. Macdonald, R. Marguerie, U. Certa, S. C. Stearns, D. B. Golstein, & L. Partridge; 2002; *Current Biology* 12: 712–23). Tony

Long and I supplied a commentary on this article: "Ageing: The many-headed monster" (M. R. Rose & A. D. Long; 2002; *Current Biology* 12: R311–12).

The paper on the number of nematode genes is "Transcriptional profile of aging in *C. elegans*" (J. Lund, P. Tedesco, K. Duke, J. Wang, S. K. Kim, & T. E. Johnson; 2002; *Current Biology* 12: 1566–572).

Chapter 13: Woody Allen and Superman

One of the many publications of the New England Centenarian Study is *Living to 100: Lessons in Living to Your Maximum Potential at Any Age* by Thomas T. Perls, Margery Hutter Silver, John F. Lauerman (1999; Perseus). There is a lot out there from this groundbreaking study.

The paradigm-shattering papers were "Slowing of mortality rates at older ages in large med-fly cohorts"(J. R. Carey, P. Liedo, D. Orozco, & J. W. Vaupel; 1992; *Science* 258: 457–61) and "Demography of genotypes: Failure of the limited life span paradigm in *Drosophila melanogaster*" (J. W. Curtsinger, H. H. Fukui, D. R. Townsend, & J. W. Vaupel; 1992; *Science* 258: 461–63).

Our attempt to evade cognitive dissonance was a letter to the editor of *Science* (T. J. Nusbaum, J. L. Graves, L. D. Mueller, & M. R. Rose; 1993; *Science* 260: 1567).

Our publication indicating the inevitability of a late-life cessation of aging was "Evolutionary theory predicts late-life mortality plateaus" (L. D. Mueller & M. R. Rose; 1996; *Proceedings of the National Academy of Science, USA* 93: 15249–253). Brian Charlesworth and Linda Partridge also supplied an interesting commentary on this sort of theoretical work: "Ageing: Leveling of the Grim Reaper" (1997; *Current Biology* 7: R440–R442).

The idea of demographic heterogeneity has been particularly promoted by James W. Vaupel, first in theory (J. W. Vaupel, K. G. Manton, & E. Stallard; 1979; "The impact of heterogeneity in individual frailty on the dynamics of mortality"; *Demography* 16: 439–54), and then in his interpretation of demographic data (J. W. Vaupel, J. R. Carey, K. Christensen, T. E. Johnson, A. I. Yashin, N. V. Holm, I. A. Iachine, V. Kannisto, A. A. Khazaeli, P. Liedo, V. D. Longo, Y. Zeng, K. G. Manton, & J. W. Curtsinger; 1998; "Biodemographic trajectories of longevity"; *Science* 280: 855–60).

Mark Drapeau's work on stress resistance and demographic heterogeneity is summarized in "Testing the heterogeneity theory of late-life mortality plateaus by using cohorts of *Drosophila melanogaster*" (M. D. Drapeau, E. K. Gass, M. D. Simison, L. D. Mueller, & M. R. Rose; 2000; *Experimental Gerontology* 35: 71–84). Larry Mueller's devastating analysis of the heterogeneity theory was published as "Statistical tests of demographic heterogeneity theories" (L. D. Mueller, M. D. Drapeau, C. S. Adams, C. W. Hammerle, K. M. Doyal, A. J. Jazayeri, T. Ly, S. A. Beguwala, A. R. Mamidi, & M. R. Rose; 2003; *Experimental Gerontology* 38: 373–86).

Some additional papers from the Curtsinger lab, including data that fails to support demographic heterogeneity, are "Slowing of age-specific mortality rates in *Drosophila melanogaster*" (H. H. Fukui, L. Xiu, & J. W. Curtsinger; 1993; *Experimental Gerontology* 28: 585–99), "Deceleration of age-specific mortality rates in chromosomal homozygotes and heterozygotes of *Drosophila melanogaster*" (H. H. Fukui, L. Ackart, & J. W. Curtsinger; 1996; *Experimental Gerontology* 31: 517–31), "Stress experiments as a means of investigating age-specific mortality in *Drosophila melanogaster*" (A. A. Khazaeli, L. Xiu, & J. W. Curtsinger; 1995; *Experimental Gerontology* 30: 177–84), and "The fractionation experiment: reducing heterogeneity to investigate age-specific mortality in *Drosophila*" (A. A. Khazaeli, S. D. Pletcher, & J. W. Curtsinger; 1998; *Mechanisms of Ageing and Development* 105: 301–17).

Our dreadnought paper on the experimental evolution of late life is "Evolution of late-life mortality in *Drosophila melanogaster*" (M. R. Rose, M. D. Drapeau, P. G. Yazdi, K. H. Shah, D. B. Moise, R. R. Thakar, C. L. Rauser, & L. D. Mueller; 2002; *Evolution* 56: 1982–991).

Our first report of the late-life fecundity plateau was "Aging, fertility, and immortality" (C. L. Rauser, L. D. Mueller, & M. R. Rose; 2003; *Experimental Gerontology* 38: 27–33).

Brian Charlesworth's latest theoretical tour de force on aging is "Patterns of age-specific means and genetic variances of morality rates predicted by the mutation-accumulation theory of ageing" (2001; *Journal of Theoretical Biology* 210:47–65). He is still in top form.

Amazingly, the late-life cessation of human aging was well documented in 1939 by M. Greenwood and J. O. Irwin ("Biostatistics of senility"; *Human Biology* 11: 1–23). They just didn't believe what they saw, a classic case of scientific discovery requiring a prepared mind.

I put forward the idea of progressively moving toward the cessation of aging in my essay for the Post and Binstock volume that arose from the Philadelphia meeting described in chapter 1.

Chapter 14: Not Even Oppenheimer

Richard Rhodes's book *The Making of the Atomic Bomb* (1987; Simon & Schuster) is the classic account of the Manhattan Project.

My first ideas concerning the postponement of human aging were published in "The evolutionary route to Methuselah" (1984; *New Scientist* 103: 15–18).

For the story of the race to sequence the human genome, remarkably written by someone who is well qualified to do so, see *Cracking the Genome: Inside the Race to Unlock Human DNA* by Kevin Davies (2002; Johns Hopkins University Press).

The 2002 article by S. J. Olshansky, L. Hayflick, & B. A. Carnes, "No truth to the Fountain of Youth" was published in *Scientific American* 286 (6): 92–95.

Malcolm Gladwell's *The Tipping Point, How Little Things Can Make a Big Difference* was published by Back Bay Books in 2000.

One of the better papers on the pathology associated with human growth hormone supplementation is "Growth hormone and sex steroid administration in healthy aged women and men: A randomized control trial" (M. R. Blackman, J. D. Sorkin, T. Münzer, M. F. Bellantoni, J. Busby-Whitehead, T. E. Stevens, J. Jayme, K. G. O'Connor, C. Christmas, J. D. Tobin, K. J. Stewart, E. Cottrell, C. St. Clair, K. M. Pabst, & S. M. Harman; 2002; *Journal of the American Medical Association* 288: 2282–292).

Chapter 15: The Long Tomorrow

My last attempt to address the challenge of postponing human aging was "Can human aging be postponed?" (1999; *Scientific American* 281 (6): 106–11). You can also get it as a *Scientific American* e-Book.

Bob Shaw's paradigmatic science fiction anti-aging novel is *One Million Tomorrows* (1970; Ace).

The leading figures in recent cell-replication research are Jerry Shay and Woodring Wright, who have shown definitively the dependence of the human Hayflick Limit on telomerase, as well as the significance of this for cancer. Some useful recent publications include "Historical claims and current interpretations of replicative aging" (W. E. Wright & J. W. Shay; 2002; *Nature Biotechnology* 20 (7): 682–88), "Telomerase and cancer" (J. W. Shay, Y. Zou, E. Hiyama, & W. E. Wright; 2001;

Human Molecular Genetics 10 (7 Special Issue SI): 677–85), and "Hayflick, his limit, and cellular ageing" (J. W. Shay & W. E. Wright; 2000; *Nature Reviews Molecular Cell Biology* 1 (1): 72–76).

Heinlein's *Methuselah's Children* was published by Signet in 1958.

Chapter 16: Travels with the Boatman

Freud paired death instincts, or thanatos, with life instincts, or eros, in *Beyond the Pleasure Principle* (1922; Hogarth Press). However, these terms are broader than the simple "death vs. sex" duality that seems intuitively obvious from the way words like eros are used in everyday culture. With the death-instinct concept, Freud was exploring the idea of a longing for quiescence, or peace, very resonant with that found in contemporary theology. Ideas like these hardly have the cogency or material significance of a scientific theory, so there is considerable room for disputes about both the meanings of thanatos and Freud's intentions in bringing the term forward, to the delight of humanists.

Index

A

Academy Awards, 51

After Many a Summer Dies the Swan (Huxley), 15

aging, 1, 11; animal species and, 57–58, 61, 63–68; caloric intake and, 76–79; cells and, 13, 18–23; cost of reproduction and, 54, 56–59, 62; death and, 135–38; demographic heterogeneity and, 103–8; dessication and, 72–77; diet and, 78–84, 86–90; force of natural selection and, 41–45; genetics and, 24–27, 33–38; genome and, 110, 115, 119, 125, 130–33; growth hormone and, 123–25; killer applications and, 121–25; mice and, 112–13, 115–16, 118–19, 123; mortality and, 103–4, 107–9; National Institute on Aging and, 49, 86, 90, 112–16; physiology and, 71–76; postponement of, 87, 90, 110–12, 118–19, 123, 129; religion and, 2–4; rodents and, 78–83; scientific respectability and, 7, 13; Smith and, 5–10, 12; stress resistance and, 71–76; synchronization of, 92–93;

theoretical science and, 5–9, 12, 19, 32–34, 45, 148; theories of, 13–18, 75; therapies for, 127–28. *See also under* evolution; specific scientists; specific topics

agricultural policies, 89

AIDS (Acquired Immune Deficiency Syndrome), 4

Alcor, 134

Alice's Adventures in Wonderland (Carroll), 60

Allen, Diogenes, 2–3, 149

Allen, Woody, 35, 99, 104, 106

Alzheimer's Disease, 127

American Association for Anti-Aging Medicine, 134

Americans, 49, 90

amyotrophic lateral sclerosis (Lou Gehrig's disease), 85

Ancient Mariner, 137

Andrews, Bill, 118

Andrews, Ralph, 117–18

anemones, 101, 102

animal species, 57–58, 61, 63–65, 68–69; patterns in, 63, 66–67

Anne of Cleves, 39

anorexia nervosa, 81, 84

Antechinus, brown (marsupial mouse), 56, 92

double agents, 17

Drapeau, Mark, 105, 158

drosophila, 19, 35, 41, 44, 94, 102; demographic heterogeneity and, 105–8; literature on, 154, 156–59; sterility in, 47–48. *See also* fruit flies; Methuselah flies

E

ecology, 9, 13, 48, 64–65

Edison, Thomas Alva, 111, 116, 118

egg production, 54, 59, 81–82. *See also drosophila*; fruit flies; Methuselah flies

Egyptian artifacts, 1, 2

Einstein, Albert, 31, 152

electricity, 110

elephants, 58

elitism, 12

elixir of life, 126–28

emotional trauma, 41

England, 28, 30, 46

environmental effects, 59, 61, 156

error catastrophe theory, 16–18, 77, 150, 151

estrogen, 124

ethical issues, 2

Eton School, 30

eugenic legislation, 57

eunuchs, 56–57

Europe, 88, 90

evolution, 68, 75, 77–78, 91–92, 103; life span and, 4

evolutionary biology, 2, 29, 52, 105, 127

evolutionary theory, 11–12, 46, 52, 74, 148, 152; cells and, 13–17; demographic heterogenity

and, 106–8; force of natural selection and, 41–45; Hayflick and, 18–23. *See also* specific theories; theoretical science

evolutionists, 7

Evolution (journal), 52

Evolution of Sex, The (Smith), 29

Ewald, Paul, 14, 150

exercise, 121

Experimental Farm, 117

Experimental Gerontology (journal), 147

experimental science, 6–9, 13–15, 34–35

experiments: design of, 36, 46, 59–61. *See also* theory

F

Faraday, Michael, 111

fat content, 73–74, 76, 78, 82–84, 123

fecundity, 36–37, 59, 66, 69–70, 82, 107–8; fertility and, 154, 159

females, 67, 154

fertility, 42, 67, 81, 123, 154, 159; reduction of, 57

Finch, Caleb, 14, 147, 150, 154, 155, 156

fish, 66

fission, 101–2

Fleming, Jim, 95

flight, 65

fly population cages, 70–72

Folk, Donna, 76

food, 70–72

Forbidden City (China), 57

force of natural selection, 33–38, 49, 52–53, 59, 73–74, 102; animal species and, 64–66; de-

Harmsen, Dolf, 10
Harrison, David, 112
Harvard University, 9–12, 23, 24, 27
Harvey, Paul, 29, 152
Hawn, Goldie, 126
Hayflick, Leonard, 2–3, 18–23, 120, 151
Hayflick Limit, 21, 23, 121–22, 151, 160
heart disease, 14, 85. *See also* cardiovascular disease
Heinlein, Robert, 133–34
hemolymph, 77
Henry VIII, 39
herbal preparations, 121
Hollywood, California, 121
honeybees, 68–69
hormonal response system, 69
hormone replacement therapy (HRT), 124–25
Horsey, Fred (grandfather), 99, 100, 136
Howell, Linda, 119
Hoyle, Fred, 8
Human Genome Project, 115
humans, 58, 108–9, 118–19, 133, 160
Huntington's Disease, 33, 151
Hutchinson, Ted, 93–94
Hutchinson-Guilford's progeria. *See* childhood progeria
Huxley, Alduous, 15, 151
hybrids, 93–94

I
ideology, 150
immortality, 19–20, 101–2, 126, 159
immune response, 18, 21

immune system, 122
inbreeding, 154
Indy fruit fly, 60. *See also Drosophila;* fruit flies; Methuselah flies
infection, 14, 86
inflammation, 150
inheritance patterns, 94, 97. *see also* genetics
insects, 21, 23, 68, 79, 92
institutional life, 30
insulin, 124, 128
Internet, 86, 121
intestines, 14, 16
invertebrates, 79, 83
in vitro cultivation, 18–21, 129
irradiation, 54. *See also* radiation damage
Itozaku, David, 87
Ives, Philip, 36

J
Jackson Laboratory, 112
Japan, 88, 90. *See also* Okinawa
Jewish Theological Seminary of America, 3
Jews, 88
Johnson, Tom, 114
Jones, Terry, 31
Journal of Gerontology (journal), 147
Judaism, 3–4
Julius Ceasar (Shakespeare), 11
Just Another Missing Kid (documentary film), 49, 51

K
killer applications, 121–25
Kingston, Ontario, 50